Andrew Goudie and John Wilkinson

THE WARM DESERT ENVIRONMENT

CAMBRIDGE UNIVERSITY PRESS
Cambridge
London New York Melbourne

Published by the Syndics of the Cambridge University Press
The Pitt Building, Trumpington Street, Cambridge CB2 1RP
Bentley House, 200 Euston Road, London NW1 2DB
32 East 57th Street, New York, NY 10022, USA
296 Beaconsfield Parade, Middle Park, Melbourne 3206, Australia

First published 1977

Typeset by Type Practitioners Ltd, Sevenoaks, Kent
Printed in Great Britain at the University Press, Cambridge

Library of Congress Cataloguing in Publication Data
Goudie, Andrew.
The warm desert environment.
(Cambridge topics in geography series)
Bibliography: p.
Includes index.
1. Deserts. 2. Arid regions. I. Wilkinson, John Craven, joint author.
II. Title. GB611.G68 910'.031'54 76-9731
ISBN 0 521 21330 4 hard covers
ISBN 0 521 29105 4 paperback

Acknowledgements
Thanks are due to the following for permission to reproduce copyright material:
Fig. 1; UNESCO, *Arid Zone Research* 28. Figs. 2,3,38; Methuen. Fig. 4; from Grigg: *The Harsh Lands,* Macmillan, London. Figs. 5,43; Armand Colin. Fig. 6; from *Geographische Zeitschrift* 58, Franz Steiner Verlag. Fig. 8; from Flint: *Glacial and Quaternary Geology,* the author and Wiley. Figs. 17,18,33,39; from Cooke and Warren: *Geomorphology in Deserts,* Batsford. Figs. 24,35; Gustav Fischer Verlag. Figs. 25,26; Institute Géographique National, Paris. Fig. 31; from Twidale: *Geomorphology with Special Reference to Australia,* Nelson, Australia. Fig. 45; CDU, Paris. Fig. 47; Geografiska Annaler 54. Figs. 48,49; from Strahler: *Introduction to Physical Geography,* the author and Wiley. Fig. 51; Lester King. Figs. 54,55,56; after Oberlander: *Physical Geography Today,* Random House. Fig. 59; from Hurst: *The Nile,* Constable. Fig. 60; South Australian Department of Mines. Fig 61; Egyptian General Development Organisation.

The maps and diagrams are drawn by Reginald and Marjorie Piggott.

CONTENTS

PREFACE

In this short book we have not tried to cover all aspects of deserts. We have consciously restricted ourselves to a consideration of warm deserts, and within that limitation we have concentrated on the physical environment and its impact on traditional methods of land utilisation. Deserts offer many excellent examples of the inter-relationships between man and his natural environment, with water availability and quality being of particular importance.

We have also attempted to summarise some of the recent developments that have taken place in the study of deserts, for although much work on the arid zones is still of a reconnaissance nature, there has in the post-war era been an increased volume of sound scientific work which has clarified many of the basic problems. Many of the traditional explanations of such phenomena as dunes, desert pavements, rock weathering and climatic fluctuations, have changed, and we have attempted to draw attention to such developments.

In a sense this is a book on desert problems. For this reason we have written at relative length on certain factors which make the world's dry zones particularly inhospitable. Thus we stress that deserts suffer from precipitation which is variable both in time and space, that they have certain inhospitable surface types (such as crusts and desert pavement), that they have soils which are frequently saline and poorly structured, that they provide a low degree of biological productivity, that they suffer from sometimes intense erosion by wind and water, and that their water supply is often saline and inadequate, or practically non-existent.

The deserts of the world vary very considerably and so we give examples of features from a wide range of places, though because of our own field experience we have paid particular attention to the deserts of the Old World, notably the Middle East, India and Africa. The distribution of deserts also partly accounts for the unnecessarily complex terminology which has been used to describe their morphology. Alas, confusion is worse compounded because field workers have frequently not understood the local terms they have borrowed (sometimes because of a different appreciation of the terrain) and because the words may be spelt in any number of ways according to the transliteration system employed. Until some bold spirit manages to establish a desert esperanto we are beset with the present verbal chaos. We, for our part, have tried to limit the number of terms employed in this book.

We are indebted to A.T. Grove for his comments and to the series editors for their advice.

Holywell, Oxford

A.S.G.

J.C.W.

PART ONE · PHYSICAL CHARACTERISTICS

1. Climate

Introduction

One of the characteristics of deserts is their variability. They are found in low and in high latitudes, in the interiors of continents and on the coasts, at high altitudes and at sea level (Fig. 1). Cold deserts, those occurring at high altitudes and high latitudes, have many of their own problems and characteristics. This book is concerned predominantly with the nature of those warm deserts which occur at relatively low altitudes in the tropics and subtropics.

Even within this category, however, there are vast variations produced by differences in climate, vegetation, surface materials and other natural factors. Moreover, the imprint of man varies. Some desert areas are moderately heavily populated while others are largely unoccupied.

Deserts are dry places and much of the variety of warm deserts reflects sometimes quite minor variations in the availability of moisture. Human activities, vegetation characteristics, animal distributions and surface processes are all to varying degrees linked with moisture availability. So any definition of a desert involves degrees of aridity, or degrees of moisture availability. One of the problems is to devise ways of expressing these differences quantitatively.

Water balance

The modern systems used for defining aridity are all based on the concept of *water balance*, that is the relationships that exist in a given area between the input of water in the form of precipitation (P), the losses due to evaporation and transpiration by plants (evapotranspiration; Et), and any changes in storage (soil moisture, groundwater, lakes, streams, etc.). In arid zones there is, by definition, an overall deficit in a year, and the size of that deficit determines the degree of aridity. Since the actual amount of evapotranspiration that occurs (AEt) will vary with many factors, not least whether there is in fact any water to evaporate, the evapotranspiration which would occur from a standardised surface never short of water is estimated instead (potential evapotranspiration; PEt). This is roughly the same as the amount of water used when growing a grass-like crop under irrigation: the volume will vary according to four climatic features; radiation, humidity, temperature and wind.

One of the first people to work on water balance measurement was an American, C.W. Thornthwaite, who devised (in 1948) a useful general aridity index which, due to its simplicity, is still much used by climatologists today, even though his approximations of the climatic variables which determine PEt are far too crude to be of much practical use to engineers and agriculturalists. According to his index:

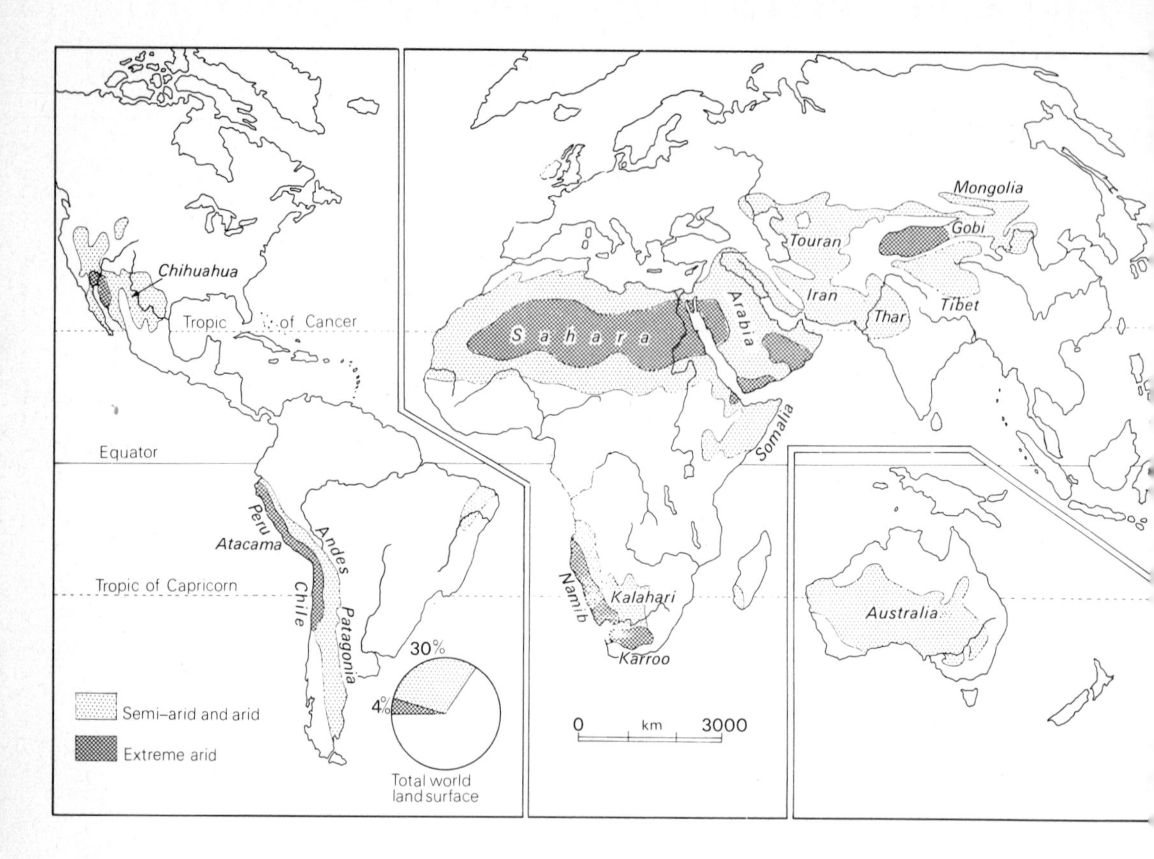

1 The distribution of world deserts according to the divisions of Meigs.

When P = PEt throughout the year the index is 0
When P = 0 throughout the year the index is – 100
(When P greatly exceeds PEt throughout the year the index is +100.)

Climates with index values between 0 and –20 are considered subhumid, between –20 and –40 semi-arid, and below –40 arid. Meigs accepted this division but further divided the last category into arid and extreme arid, with extreme aridity being defined as the condition experienced in any given locality in which at least twelve consecutive months without any rainfall have been recorded, and in which there is not a regular seasonal rhythm of rainfall. This threefold division of aridity is used in the construction of the map of world desert areas shown in Fig. 1.

The extent of aridity as defined

Under the definition adopted, the proportion of the surface of the earth covered by arid conditions is roughly one-third, with approximately 4% of the land surface extremely arid, 15% arid, and about 14.6% semi-arid. In all about 49 million square kilometres out of the world's total land area of 135 million square kilometres are affected.

Basically the dry climates of the continents occur in five great provinces separated from one another either by oceans or by the wet equatorial zones. Of these five provinces the largest, the North Africa–Eurasia province, is larger than all the remaining dry areas of the world combined. It includes the Sahara and a series of other

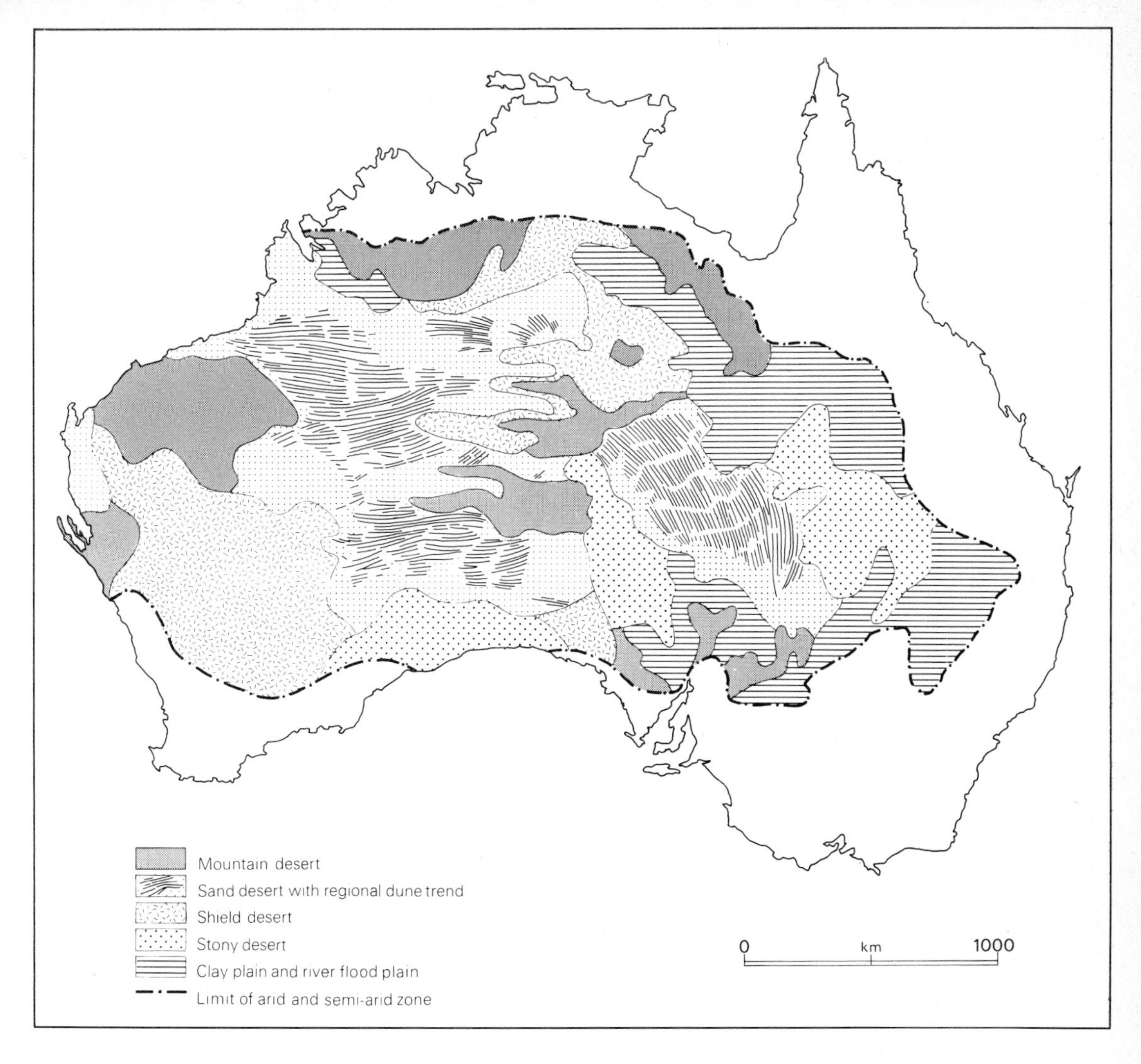

2 Map of the main desert surface types in Australia.

deserts extending eastwards through Arabia to Pakistan and India and Central Asia.

The Southern African province consists of two main parts: the coastal Namib Desert which occurs along the Atlantic coast of South Africa, South West Africa (Namibia) and Angola; and the Kalahari and Karroo inland dry zones.

The South American dry zone is confined to a strip along the west coast (the Atacama) and on the east side of the continent toward the southern portion of the continent (the Patagonian Desert).

The North American dry province has a variety of types and occupies much of Mexico and the south-west United States.

The fifth province is that of Australia, where much of the central and western parts of the island is occupied by various types of desert features (see Fig. 2), though there is no extreme aridity. As can be seen, desert surface types are numerous.

The causes of aridity

Many of the world's warm deserts occur in zones where there is subsiding air, relative atmospheric stability, and divergent air flows at low altitudes, associated with the presence of great high pressure cells around latitude 30°. Such areas are only occasionally subjected to incursions of precipitation-bearing disturbances and depressions such as characterise both the convergence zone of the tropics (the Intertropical Convergence Zone) and the belts of circumpolar westerlies of high and middle latitudes. The trade winds which blow across these zones are evaporating winds and are not favourable to precipitation inasmuch as they become warmer as they move from higher to lower latitudes.

Reinforcing these global tendencies are the effects of more local factors. Of these, continentality, or distance from the sea, is a dominant one, and plays a large part in the location and characteristics of, for example, the mid-latitude deserts of Central Asia. This situation can also be illustrated from Australia where the Great Divide and other mountains on the east side of the island, lie athwart the direction of the prevailing south-easterly trade winds; this creates a 'rain shadow' effect which accentuates the aridity of the great 'dead heart' of Australia. The rain-shadow effect produced by great mountains can create arid areas in the lee of the mountains even when continentality is not particularly marked, as in the southern Deccan of India or in Patagonia where the Western Ghats and the Andes respectively play the role. Conversely, within dry regions, uplands like the Hoggar and Tibesti of the Sahara sometimes cause sufficient lift for orographic rain to provide less arid conditions.

Some deserts are specifically coastal, the Atacama of western South Africa, and the Namib of South West Africa (Namibia) and Angola being good examples. In these situations any winds that do blow onshore tend to do so across cold currents produced by movement of water from high latitudes to low and associated with the upwelling of cold waters from the ocean depths. Cold or cool winds have a relatively small moisture-bearing capacity and when warmed during their passage over the land they become stable and thereby reinforce the stability produced by the global stability of these latitudes referred to at the beginning of this section.

Thus the subtropical highs are the major cause of aridity. However, the arid and semi-arid zones are not always continuous around the Earth at 30° S and N. For example, in Asia the rain-bringing summer monsoon gives rain in considerable quantities in parts of northern India. In other parts of the belt the high-pressure cells are disrupted into a series of local cells, notably over the oceans, where air moving clockwise around the equatorial side of the cells brings moist air to the eastern margins of the continents in the Caribbean, Brazil and East Africa.

It is, however, necessary to appreciate that by creating changes in moisture conditions man may himself have created certain of the arid characteristics of marginal areas. For example, through deforestation and farming man has in places encouraged water to run off rather than to infiltrate, he has caused gullying which may lower groundwater levels, and he has stripped large areas of soil cover so that conditions for vegetation growth are rendered less favourable. This man-induced development or extension of deserts is a severe problem, and is often called *desertisation.*

able 1. Extremes of shade temperature in deserts

ocation	Extreme maximum (°C)
nsalah (Libyan desert)	54
eath Valley (California)	57
zizia (Tripolitania, Libya)	58
indouf (Algeria)	57
/illiam Creek (Australia)	48.3

able 2. Ground surface temperature maxima

ocation	Temperature (°C)
ahara	78
alahari	72
ucson (Arizona)	71.5
alifornia	70
gra (India)	69
ed Sea Hills	82.5

he nature of desert climates in the tropics and subtropics

)eserts in the tropics and subtropics are basically characterised by igh temperatures, a great excess of potential evaporation over recipitation, and a high variability of precipitation totals, distribu-ion and intensity. Because the ground surface is often dry and elatively clear of vegetation cover, wind plays a more prominent ole than in many other zones of the earth's surface. Evaporation ates are generally great, often fifteen to twenty times the annual recipitation, and this is caused by high temperatures, high wind elocities (in some localities) and slight cloud cover. Clear skies revail over 70% of the time, and even longer in the summer months. Iumidity tends to be low, with 15–30% being characteristic of most nland deserts, though values as low as 5% have been recorded in arts of the Sahara. However, in coastal deserts, such as the Namib nd the Atacama, relative humidity may be 100% and fogs may be requent.

emperature

n terms of temperature the distinction between coastal and interior eserts is an important one, for while coastal deserts tend to have elatively low seasonal and diurnal ranges of temperature, interior eserts can be subjected to extremes which are not equalled in any ther climatic region.

The sheer extremes of temperature recorded at certain interior tations are listed in Table 1. These are of interest in themselves, but nore important in terms of human and animal comfort is the fact hat temperatures in excess of 37°C may occur for many days on nd in the summer months without respite. Because of the clear ight skies there may be a marked reduction of temperature at ight, and daily ranges of 17 to 22 deg C are normal, though in)eath Valley, California, a maximum diurnal range of 41 deg C was ecorded in August 1891, and the mean diurnal range for that nonth was 35 deg C. Ground surface temperatures show even

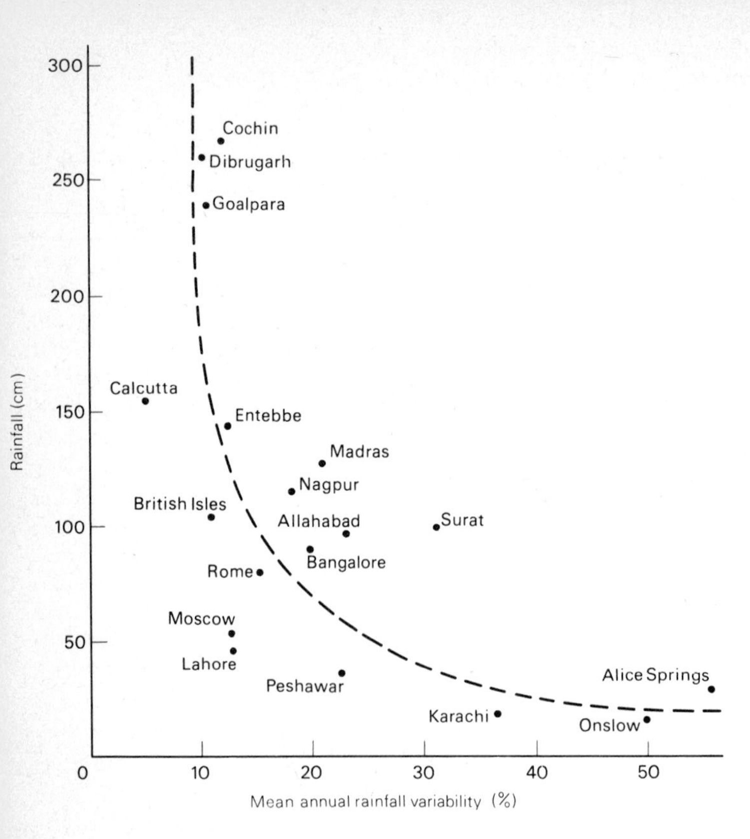

3 The relationship between mean annual rainfall variability and mean annual rainfall.

greater ranges with consequences for rock weathering which will be discussed in a later section. Sand, soil and rock have been recorded as reaching temperatures as high as 82°C (Table 2).

The modification of the temperatures of the coastal deserts along the western coasts of the continents by the presence of cold currents and upwelling water means that temperature ranges over the year are low, so that Walvis Bay, in South West Africa (Namibia), for example, has a mean annual monthly range of only 6 deg C, and Callao in Peru has an annual range of only 5 deg C. Daily ranges are also low, generally only about 11 deg C, or about half what one would expect to experience in the interior of the Sahara, and annual values are also generally moderate, with the average annual temperatures in the Atacama being around 19°C and those at Walvis Bay in the Namib being 17°C.

Precipitation variability in time and space

One important feature of desert climatology is the relatively high variability of the rainfall through time. Variability can be expressed as a simple index:

$$\text{Variability (\%)} = \frac{\text{the mean deviation from the average}}{\text{the average}} \times 100.$$

Fig. 3 illustrates values of variability plotted against mean annual rainfall for selected stations. Desert stations like Alice Springs, from the so-called Dead Heart of Australia, show up clearly as having particularly high variability. European locations like Rome have a precipitation variability of only about 14%, whereas a variability of 80–100% occurs in the central Sahara, more than 100% in the Libyan Sahara, and even 150% at Dakhla (between Koufra and Kharga).

4 World map of rainfall variability. The meaning of variability is explained on p. 6. Note the way in which the major desert belts show variability that is often in excess of 30%.

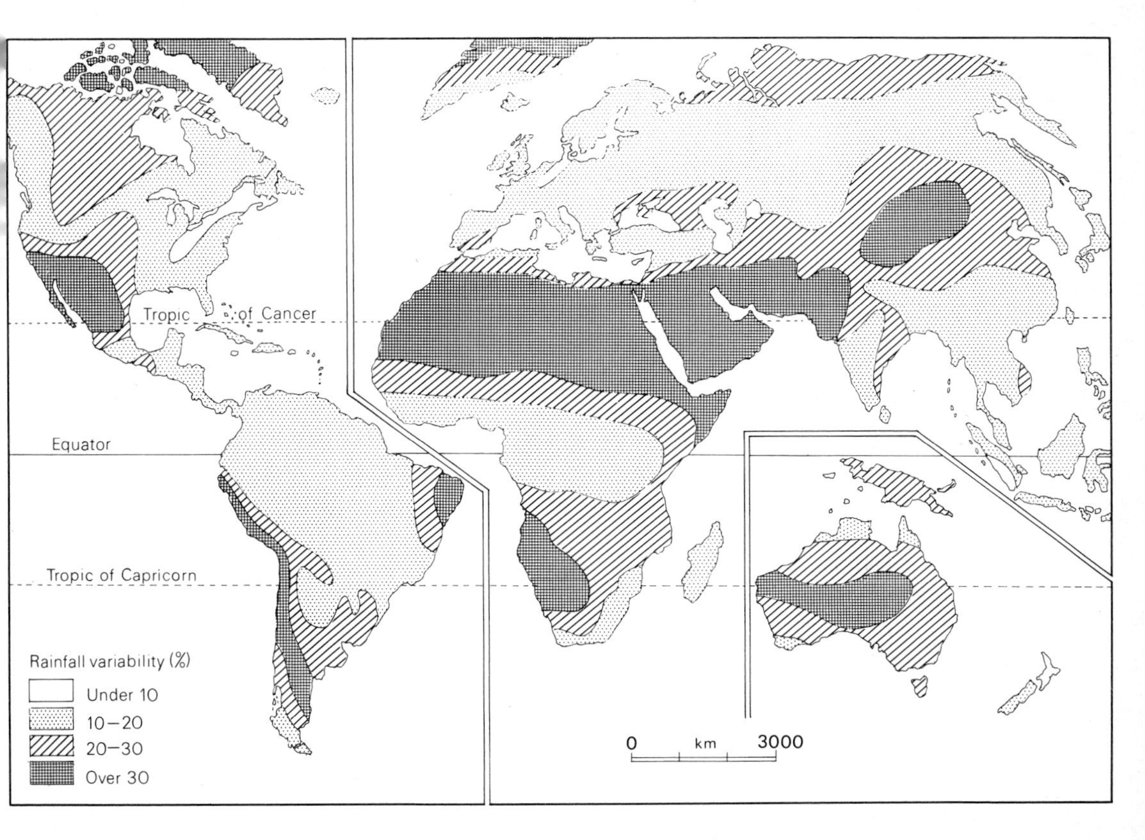

This pattern is further clarified in Fig. 4, which is a world map of rainfall variability. It is useful to compare this map with the map of world deserts in Fig. 1. The correspondence is striking.

Precipitation in arid zones, in addition to showing temporal variability, also shows considerable spatial variability. Indeed, desert rainfall is often described as being 'spotty'. This characteristic is shown clearly in Table 3 where twenty rain gauges within an area of only 10 hectares (about 25 acres) in the Negev of Israel, had catches ranging from 2.2 to 7.8 mm in the same storm. In other words, even within a very small area rainfall in some places was over three times

Table 3. Rainfall data, Avdat, Negev, Israel, 1 March 1960

	Rain gauge	Rainfall (mm)	Rain gauge	Rainfall (mm)
Automatic	1	7.3	8	4.4
Automatic	2	3.4	9	2.4
Standard		7.8		
Small	1	7.0	10	2.9
	2	6.8	11	4.2
	3	7.3	12	3.0
	4	7.0	13	4.8
	5	5.6	14	5.8
	6	2.2	15	5.0
	7	4.4	16	5.5
			17	4.6

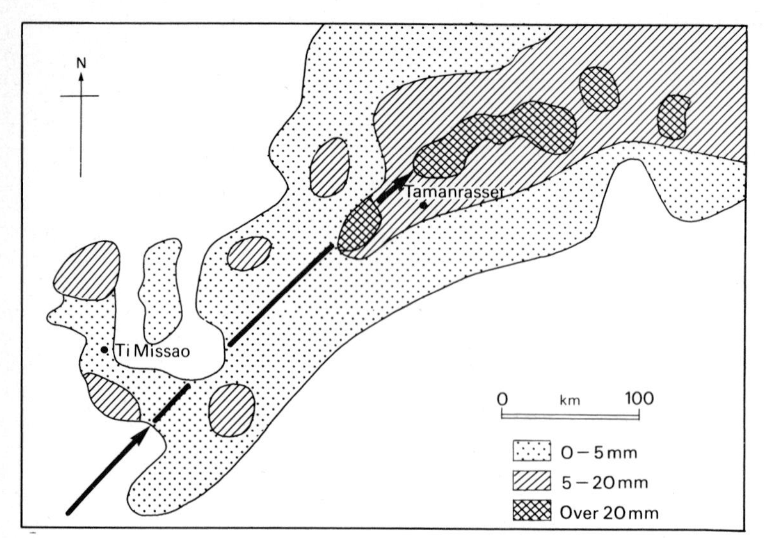

5 Precipitation spottiness in one storm near Tamanrasset, in the Sahara Desert. The arrow shows the track of the storm.

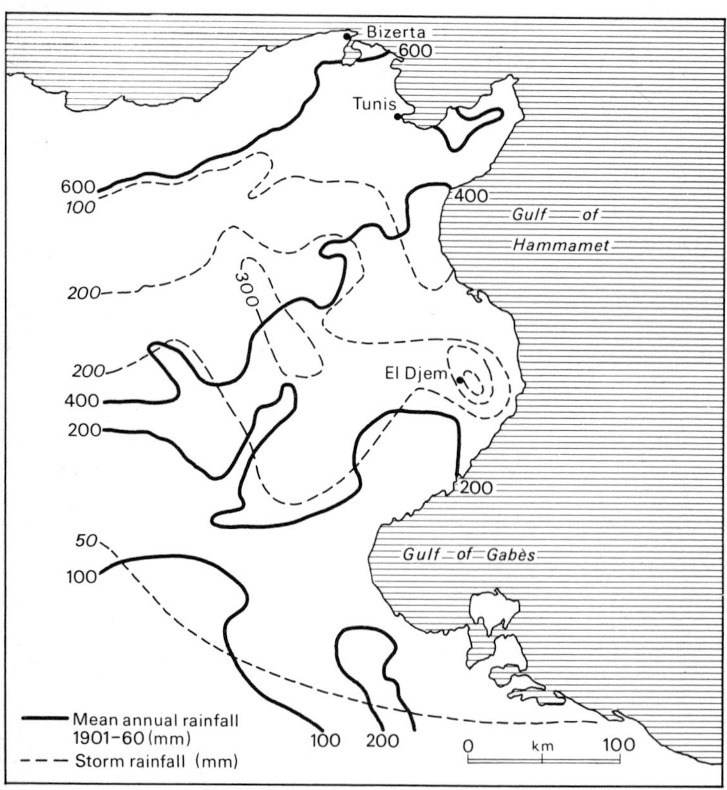

6 The distribution of rainfall during the Tunisian flood of 25–27 September 1969.

that in others. This is further brought out in Fig. 5 which is for a storm in the Sahara near Tamanrasset. Certain discontinuous zones of relatively high rainfall show up clearly.

One feature of the high precipitation variability of deserts is that sometimes very high rainfall totals are recorded in individual storms, the effects of which may be striking. Table 4 gives a list of some such events for various stations. Fig. 6 illustrates the nature of one particular storm, that of September 1969 in Tunisia, and relates the

Table 4. Extremes of precipitation

Location	Date	Mean annual precipitation (mm)	Storm precipitation
Chicama (Peru)	1925	4	394 mm
Aozou (Central Sahara)	May 1934	30	370 mm/ 3 days
Swakopmund (South West Africa; Namibia)	1934	15	50 mm
Lima (Peru)	1925	46	1524 mm
Sharja (Trucial Coast)	1957	107	74 mm/ 50 mins
Tamanrasset (Central Sahara)	Sept. 1950	27	44 mm/ 3 hours
Bisra (Algeria)	Sept. 1969	148	210 mm/ 2 days
El Djem (Tunisia)	Sept. 1969	275	319 mm/ 3 days

falls in that storm to the mean annual rainfalls in that country. The spectacularly high falls around El Djem are especially notable.

This phenomenon is emphasised further in Table 5 where a series of African stations shows that in deserts maximum falls in 24 hours may exceed the long-term annual precipitation values, in some cases by a marked amount. Likewise, in Fig. 7, the relationship between mean annual rainfall and the maximum recorded precipitation in 24 hours in southern Africa is demonstrated.

Table 5. Maximum daily falls of precipitation in African desert stations

	No. of years of record	Mean annual precipitation (mm)	Max. annual precipitation (mm)	Min. annual precipitation (mm)	Max. precipitation in 24 hours (mm)
Khartoum (Sudan)	30	164	382	76	80
Faya-Largeau (Chad)	30	17	48	tr	48
Nouakchott (Mauritania)	25	156	–	–	249
Bilma (Niger)	27	22	–	0	49
Dongola (Sudan)	30	23	60	0	36
Atar (Mauritania)	33	106	–	–	69
Etienne (Mauritania)	33	27	–	0	83
Wadi Halfa (Sudan)	24	3	33	0	19
Dakhla (Egypt)	25	0.5	11	tr	8
Quseir (Egypt)	25	4	34	tr	20
Cairo (Egypt)	25	24	63	3	44
Sollum (Egypt)	20	95	324	4	121
Galcaio (Somalia)	26	149	448	33	160
Berbera (Somalia)	40	49	178	2	132
Djibouti (Horn of Africa)	64	129	300	10	211
Luderitz (South West Africa; Namibia)	20	18	59	1	31

–, not available.
tr, trace.

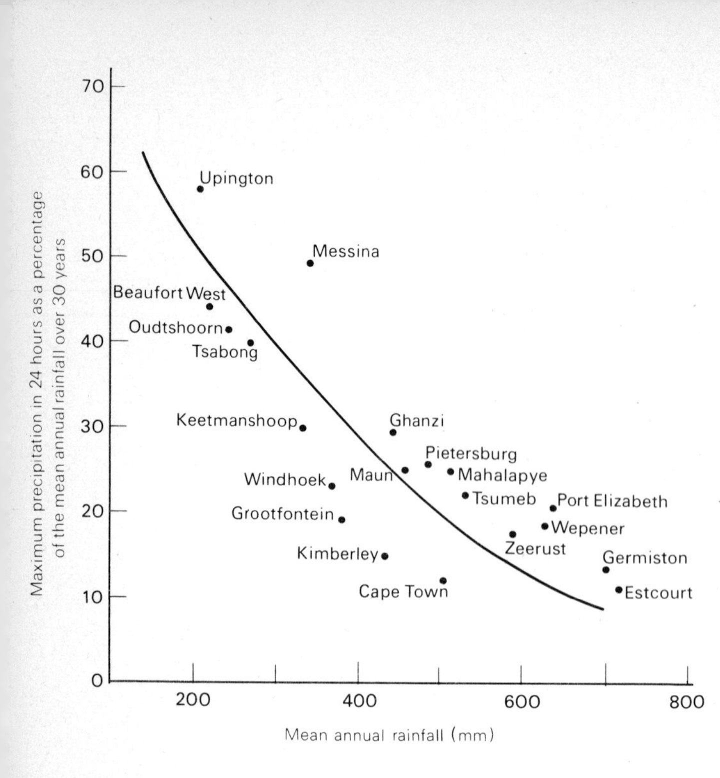

7 In Southern Africa, maximum daily rainfalls in arid areas tend to form a higher percentage of the mean annual rainfall than in more humid regions.

Precipitation in coastal deserts

Extremely low rainfall totals are characteristic of many of the great coastal deserts. For example, mean annual totals in millimetres are 30 for Callao (Peru), 15 for Swakopmund (South West Africa; Namibia) and 35 for Port Etienne in Mauritania.

The rainfall totals, however, whilst extremely meagre, do not tell the whole story. On the one hand, even these totals over a period of decades are sometimes accounted for by one extremely large storm, while on the other the *rainfall* records may underestimate the amount of total *precipitation* occurring, for the coastal deserts are greatly affected by wetting fogs which may deposit more moisture than does true rainfall. Thus at Swakopmund and Walvis Bay in South West Africa (Namibia) the mean annual fog precipitation may be 35–45 mm and fog may occur on 200 days in the year, and extend as much as 110 km inland. In the Peruvian desert there is a similar situation and in the period June to October the area is greatly affected by the strong development of low cloud called *Garua*, which furnishes enough moisture to support a growth of vegetation on any hills or mountains within its reach near the coast (see Chapter 6).

Hot winds

In many desert areas extremely desiccating conditions which cause great human discomfort are created by the onset of hot, dry, dusty winds of interior origin. In the province of Spanish Sahara, for example, the normal wind is the *Alisio* which blows from the north and brings cool conditions. However, occasionally the *Irifi*, a dreaded easterly wind, blows and brings dust and heat from the Sahara. In March 1941 the Irifi raised the temperature at Smara from 18.3°C at noon to 42.8°C at 4 p.m. and kept it above 39.4°C for 36 hours. The equivalent wind further north on the Moroccan coast is the *Chergui*, while in Libya it is the *Ghibli*, and in other

Table 6. The precipitation levels on the humid margin of modern continental dune fields contrasted with the current precipitation levels on the humid margins of fossil dune fields developed during arid phases of late Pleistocene times.

Location	Current precipitation limit of moving dunes (mm)	Precipitation limit of fossil dunes today	Shift (km)
Arizona	238–254	305–380	–
W. Africa	150	750–1000	–
Rhodesia (Zimbabwe)	300	*c.* 500	–
Australia	100	–	900
India	200–275	850	350
Sudan	–	–	200–450
Texas	–	–	350
Southern Kalahari	175	650	–

parts of the Mediterranean the *Scirocco*. On the south side of the Sahara there is the *Harmattan* and on the north side the *Khamsin*.

Climatic change in deserts

In addition to the year-to-year variability of climate which has already been described, rather longer term fluctuations and changes have taken place in most desert areas. Such changes have been important influences on human history and migrations and have greatly affected the nature of soils and landforms in present-day desert areas. Only a very few desert core areas may have escaped the changes in rainfall and temperature which took place at the same time as the Pleistocene ice sheets were waxing and waning in high latitudes.

Many deserts have been subjected to increased rainfall in the past, and such periods are called *pluvials* or lacustral phases. Many deserts, by contrast, have also been subjected to even greater aridity or *interpluvials*. In most deserts there is evidence for both phases having occurred at least once.

The presence of greater precipitation in the past is inferred from various lines of evidence: high lake levels in closed basins; expanses of fossil soils of humid type, including deep weathering products such as duricrusts (see p. 16); great spreads of spring-deposited lime, called *tufa*, indicating greater spring activity; vast river systems which are currently inactive and blocked by dune fields; and animal and plant remains together with evidence of former human habitation in areas now too dry for man to survive.

The evidence for reduced precipitation tends to be based on a smaller range of data, but in many areas there is good reason for believing that the great sand dune fields were formerly more extensive than they are at present, extending into areas which are now wooded. Data for the former extensions of such dune fields are given in Table 6. Active sand dunes cannot form in inland situations where there is anything like a continuous vegetation cover, so that their presence in currently wooded areas and in areas with precipitation above 150–250 mm per year indicates that they have become fossil features in the landscape. Such fossil dunes are often deeply weathered, gullied by water activity, and overlain by stone tools some thousands of years old.

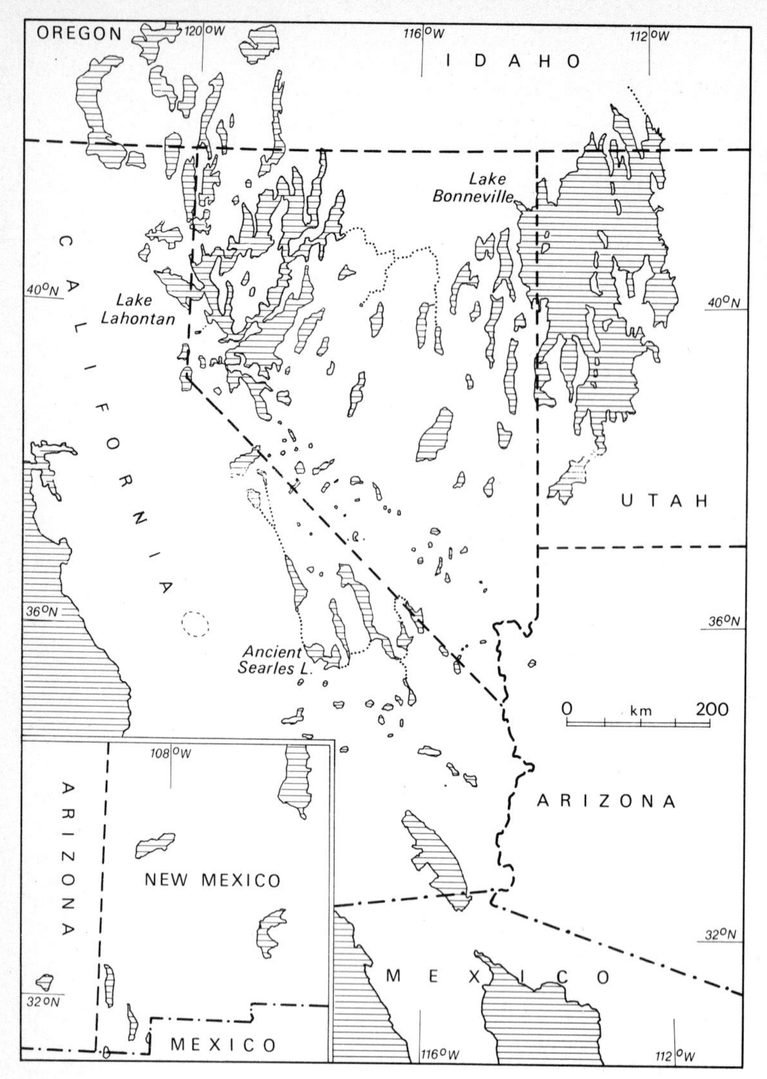

8 The dimensions of pluvial lakes in the western United States during late Pleistocene dry phases. Note in particular the great size of Lake Bonneville. At the present time the lakes are much less extensive and some are non-existent.

A good example of the complex alternations of humid and arid conditions during the Pleistocene is provided by a study of the Sahara and its margins. Lake Chad, for example, possibly stood formerly at 380–400 m above sea level (compared with its present height of about 280 m) around perhaps 55,000 years ago, but later drier conditions occurred and big dune fields were established, especially between latitudes 11° and 20° N. Subsequently humid conditions returned and a lake rising to a level of about 320 m occupied the centre of the Chad Basin (probably about 10,000 years ago), and some flooding of dune fields took place. Grove has named this lake 'Mega-Chad'. It seems likely that a lake stood at the 320 m level until as late as 5000 years ago.

Some of the pluvial lake basins reached very great dimensions. Lake Bonneville (Fig. 8) in the south-west USA, for example, which now has a water area of 2600–6500 km^2, in late Pleistocene dry phases had an area of 51,700 km^2 (almost the size of present-day Lake Michigan). There was another colossal lake in the area of the present Aral and Caspian seas. It covered an area of over 1,100,000 km^2 and extended 1300 km up the Volga river from its present mouth. Similarly, the Dead Sea, Lake Tiberias and the now-drained Lake Hula were at one time united as the 'Lisan Lake'. This lake

able 7. Five-year running mean percentage of normal rainfall*
ntred on 1957 and 1970

ocation	1957	1970
kaner (India)	114	71
dhpur (India)	115	68
hartoum (Sudan)	122	80
gades (Niger)	130	44
essalit (Mali)	140	63
ao (Mali)	114	75
ouakchott (Mauritania)	106	74
tar (Mauritania)	121	52
ean	120	66

A five-year running mean is computed by taking the mean value for the first set of five values in a series, then removing the first of the five and substituting the sixth, taking the mean of that set of five values, and so forth.

xtended from the south shore of Lake Tiberias to 35 km south of e south shore of today's Dead Sea. In all, the water volume must ave been 325 km^3 compared to the 136 km^3 of the present Dead ea.

Although climatic change of a marked degree appears to have ken place in the Pleistocene, it is apparent from the study of eteorological records dating back in some cases to the middle of e last century, that appreciable fluctuations may still take place. hus, for example, the extreme soil erosion and dust blowing of the Dust Bowl' years of the 1930s in the dry western parts of the nited States corresponded in time to a period of greatly reduced infall and of higher than average temperatures. Similarly, after oout 1895 a decline in precipitation took place in large parts of the alahari, central Australia and north-western India. Recent studies ave shown that for a great belt extending from Mauritania, through thiopia across to parts of the Thar Desert in northern India, rain-ll during the years 1957 to 1971 has been only between 60 and 0% of the long-term mean, causing great drought and hardship in rge parts of West Africa and Ethiopia. The change in the situation om the relatively favourable years of the mid 1950s to the isastrous years of the late 1960s and early 1970s is shown in able 7.

2

Surface materials

Introduction

Although the climatic conditions discussed in Chapter 1 have the greatest significance for the development of desert landscapes and for man's occupation of arid lands, deserts also possess other characteristics which need considerable attention. One of the most important of these is the nature of the surface materials on which geomorphic processes operate and man conducts most of his activities. Certain of these surface materials are highly distinctive, and many are difficult to use for economic purposes. They result from the low rainfall, sparse vegetation cover (discussed in Chapter 3) and the operation of abnormal hydrological and geomorphic processes. Among the most striking surface materials are stone pavements, various types of crusts, varnished rock surfaces, coastal salt flats, angular rock debris produced by rock splitting processes, and various soils with high salt contents and low quantities of clay and organic material.

Stone pavement

Of the various surface types found in deserts stone pavement is one of the most characteristic, though like many landform types similar forms may result from a wide variety of processes (equifinality).

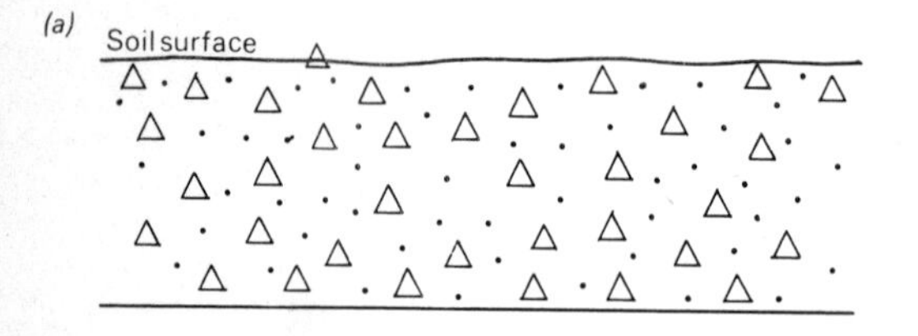

(b) Former soil surface

Vertical extent of deflation

(c)

Vertical extent of deflation

Pavement surface

9 The classic deflation model of stone pavement development, with an initial alluvial sediment containing both fine and course materials *(a)*, being subjected to deflation *(b)*, until such time as the surface is lowered to such an extent *(c)*, that a lag of coarse material is left at the surface, the fine material having been blown away.

10 Although patterned ground produced by the sorting of stones into polygons and stripes is particularly characteristic of periglacial areas, similar phenomena do exist in deserts, and are caused by the expansion and contraction of clays and salts. In the case of this example from hills on the margins of the Dead Sea, salt expanding and contracting with changes in humidity is probably the main cause of the pattern.

The true stone pavement consists of armoured surfaces composed of rock fragments, either angular or rounded, usually only one or two stones thick, set in or on matrices of finer material such as sand, silt and clay. Local names include *gibber plains* in Australia, and *hammada* and *reg* in North Africa. They are especially common where mixed and usually unsorted alluvial materials occur.

The traditional hypothesis that has been put forward to explain them is that they are a product of deflation of fine materials. This removal by the wind of fine materials from a deposit of mixed grain size would tend to leave the coarse contents as a lag or residue at the surface (Fig. 9). Vegetation cover, salt crusts, and compaction by raindrops may reduce the importance of this process in particular situations, and other hypotheses need to be put forward.

A similar result could occur as a consequence of water sorting. Coarse materials could remain at the surface after raindrops had dislodged fine materials and they had been flushed away by running water in the form of sheet flow.

However, many pavements must be explained by reference not only to lateral removal of fine materials but to the vertical movement of soil materials. For example, alternate freezing and thawing of a saturated sediment of mixed type (though rare in warm deserts) leads to stones moving to the surface. Wetting and drying can have the same effect and the alternate hydration and dehydration of salts in saline soils must also be considered.

On some desert pavements, certain strange patterns may develop as a result of humidity changes (Fig. 10). In the gibber plains of

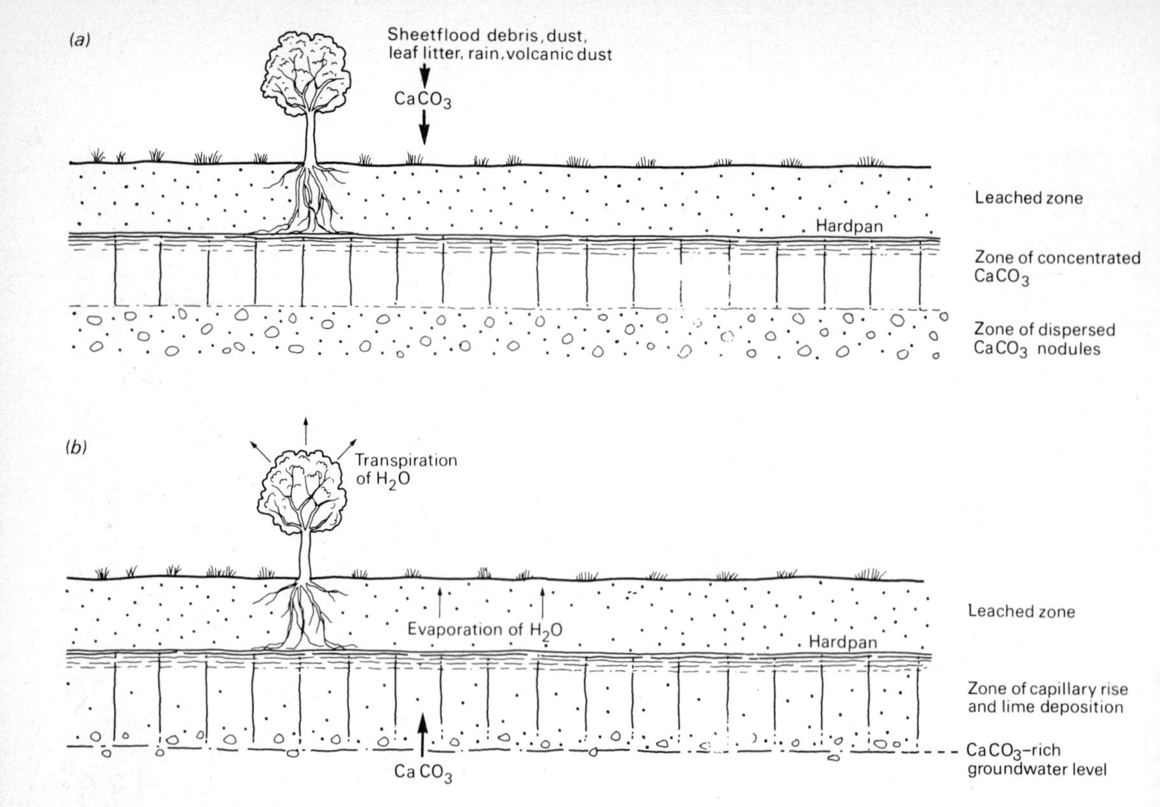

11 Two simple models of calcrete formation. *(a)* The eluviation model, in which carbonate material is leached downward from various sources, especially aeolian dust; and *(b)* the *per ascensum* model in which carbonate-rich material is deposited by upward moving solutions from groundwater.

Australia two patterns occur, called *circular gilgai* and *stepped gilgai.* The former are raised rims of pebbles surrounding a relatively bare area of clay topsoil, and they have widths of up to 3 m. They occur predominantly on flat areas whereas the stepped gilgai occur on slopes. The clay matrix of these soils expands and contracts as it is alternately moistened and dried, and wetting, therefore, tends to lead to doming which displaces pebbles sideways.

Desert crusts

Another surface type found over large parts of the arid zone is the desert crust, formed by the preferential accumulation of particular salts at or near the surface. Some of these crusts, which have been called *duricrusts,* are probably relict features of more pluvial phases, for example the laterites (iron-rich crusts) and silcretes (silica-rich crusts) associated with the extensive deep weathering layers of Central Australia. However, two particular types of crusts, the gypsum crust (*gypcrete*) and the lime crust (*calcrete*), are currently forming in many deserts. Both types of material accumulate basically because there is insufficient water activity to lead to their leaching completely out of the soil and into the rivers.

Gypsum crusts are particularly characteristic of the very driest deserts like the Namib Desert near Walvis Bay, and of low-lying areas such as occur to the south-west of Lake Eyre in Australia. The crusts, which may be quite tough, are often 30 to 100 cm thick.

Calcrete crusts may occur in areas with precipitation as high as 600 to 850 mm per annum, but are best developed in areas with precipitation less than 500 mm, where they form layers up to 50 m thick. Gradations can be seen from small accumulation of lime nodules to solid and banded hardpans which greatly impede root development of desert plants.

12 A common aspect of the flora of very dry deserts is lichens. The examples shown here, on basalt boulders in the Namib Desert, probably gain much of their moisture requirement from the frequent fogs. They may also play a role in desert varnish formation.

Many of these crusts develop through a process called *eluviation* (Fig. 11*a*) whereby salts are leached from the surface of a soil to lower horizons. Because there is insufficient moisture to leach them right through they accumulate at depth in the soil, either because of evaporation or because of other factors which cause precipitation of salts. Carbonate or gypsum may be added to the soil surface as dust derived from deflation elsewhere, by vegetation or by sheetfloods, so that thick layers can accumulate. However some crusts may form in an almost exactly opposite manner (the *per ascensum* model). They develop through the upward movement of moisture and salts from groundwater by capillary attraction (Fig. 11*b*). When the water approaches the surface, evaporation takes place and salt is deposited. Such a process, however, would tend not to take place if groundwater were deeper than about 3 m below the surface, as capillarity will normally operate only over a limited vertical range.

Desert varnish

Another crust, of somewhat different type from those so far described, is the thin patina of lacquer which covers many rock outcrops, and which is called *desert varnish* (Fig. 12). This coating, which tends to be dark red to black in colour, is composed largely of iron and manganese oxides with silica. The processes which form it are not fully understood, but capillary rise of salts under the influence of high evaporation may play a major role. The varnish may develop quickly given suitable conditions and in the Mojave

13 The sabkha between Abu Dhabi and the main desert platform is normally dry and hard enough to support the weight of light vehicles. After rainfall or high tides, however, the water may lie for weeks and vehicles may become bogged down and break through the surface crust.

desert in the American south-west a good varnish was seen to develop in only 25 years. However, in most places the speed of varnish formation is probably very much slower.

Marine sabkhas

One particular type of surface found in coastal areas of North Africa and Arabia is the *sabkha.* These are low-lying salt flats which are periodically covered by high tides or by rainstorms (Fig. 13). They occur just above normal high-tide level at about 0.5–3.0 m. In addition they have high water tables, a depth of only 0.03–1.0 m being common. Because of the frequency of inundation, the high water table levels and the very high summer water temperatures (often over 35°C for some weeks in summer), large spreads of salt, called *evaporites,* precipitate out. These include anhydrite, gypsite and halite.

The flats of the Trucial Coast on the Gulf are particularly extensive, stretching over 320 km along the coast and attaining a maximum width of almost 30 km. It seems that the sabkhas have grown seawards, as a result of prolonged sedimentation in the intertidal zone since world-wide (*eustatic*) sea levels became relatively stable 6000 to 7000 years ago. This sedimentation has largely been chemical (evaporite deposition) and biogenic (resulting from the growth of calcareous organisms). Deflation and deposition appear to have been in approximate balance.

14 Salt weathering is a potent force in desert landform development. In southern Ethiopia salt accumulates around closed lake basins in the Rift Valley and this attacks volcanic rocks to give microscale bizarre weathering forms such as the miniature inselberg-and-pediment landscape illustrated here.

Rock breakdown in deserts

Over many desert surfaces there are large quantities of shattered rock debris, split boulders and individual crystals of igneous rocks. In spite of the relative absence of water, weathering seems to operate to an important degree, and some characteristic small weathering forms are common. These include scales, sometimes several metres thick, which have broken away from a core of rock, especially igneous rock, by the process called *exfoliation.* There are also large boulders which have been split cleanly down their centres to produce the so-called Kernsprung of German authors. Rocks are also often pitted by honeycomb weathering called *alveoles,* and they may be pitted by larger features called *tafoni,* which have arched entrances, convex walls, overhanging upper margins, and fairly smooth, gently sloping floors.

The relative absence of soil and vegetation and the resulting exposure of bare rock may explain why these features appear more prominent in deserts than elsewhere. One type of weathering has been held to be a special feature in deserts and it has been postulated that it helps to explain the existence of exfoliation and Kernsprung development. This type of weathering is called *insolation weathering,* and is believed to be caused by alternations of heat and cold leading to expansions and contractions in the rock so that it eventually splits. Extremes of surface diurnal temperature in deserts are considerable, and daily ranges of over 75 deg C have even been recorded. The process may be favoured by the different expansions of different

15 Because water is restricted, solution of limestones (see p. 42) is seldom sufficient in deserts to lead to the development of major solutional features. However, small weathering rills may result from fog drip and dew in coastal deserts. In this case a marble outcrop in the Namib Desert is affected.

minerals in an igneous rock. However, experimental work in the laboratory has shown that this process does not operate quickly in the absence of water and so other mechanisms of rock breakdown have been examined in recent years.

Of the other mechanisms of physical or mechanical weathering, mechanisms which do not involve actual chemical change of rock constituents, one of the most effective is salt weathering (Fig. 14). In desert areas salt is frequently present in groundwater, fogs, soils and elsewhere. When water in which salt is dissolved is evaporated, the salt crystallises out. The growth of the crystals sets up considerable forces, so that if a rock contains a solution of salt within its pore space it may be disrupted by the crystal growth consequent upon evaporation. This process is further assisted if the salt involved is one which is capable of taking up water, that is becoming *hydrated.* A hydrated salt, with much water of crystallisation, would have a greater volume than a salt without such water (a *dehydrated* salt). The passage from a dehydrated to a hydrated state, which can take place because of temperature or humidity change, could lead to the expansion of salt within a rock, and hence to the rock's disruption.

Further mechanical weathering can take place as a result of wetting and drying. Rocks like shale seem susceptible to this type of breakdown. Also, in some deserts at high altitudes, frost action may be important during the winter months.

However, it needs to be remembered that in all deserts there will from time to time be moisture present so that chemical changes will

6 Desert weathering processes can be mulated in the laboratory using simple echniques. In this experiment blocks of chalk were subjected to a variety of processes and their weight was recorded after each cycle. Certain salts seemed to be highly effective weathering agents when solutions of them crystallise out in the pore spaces of the chalk.

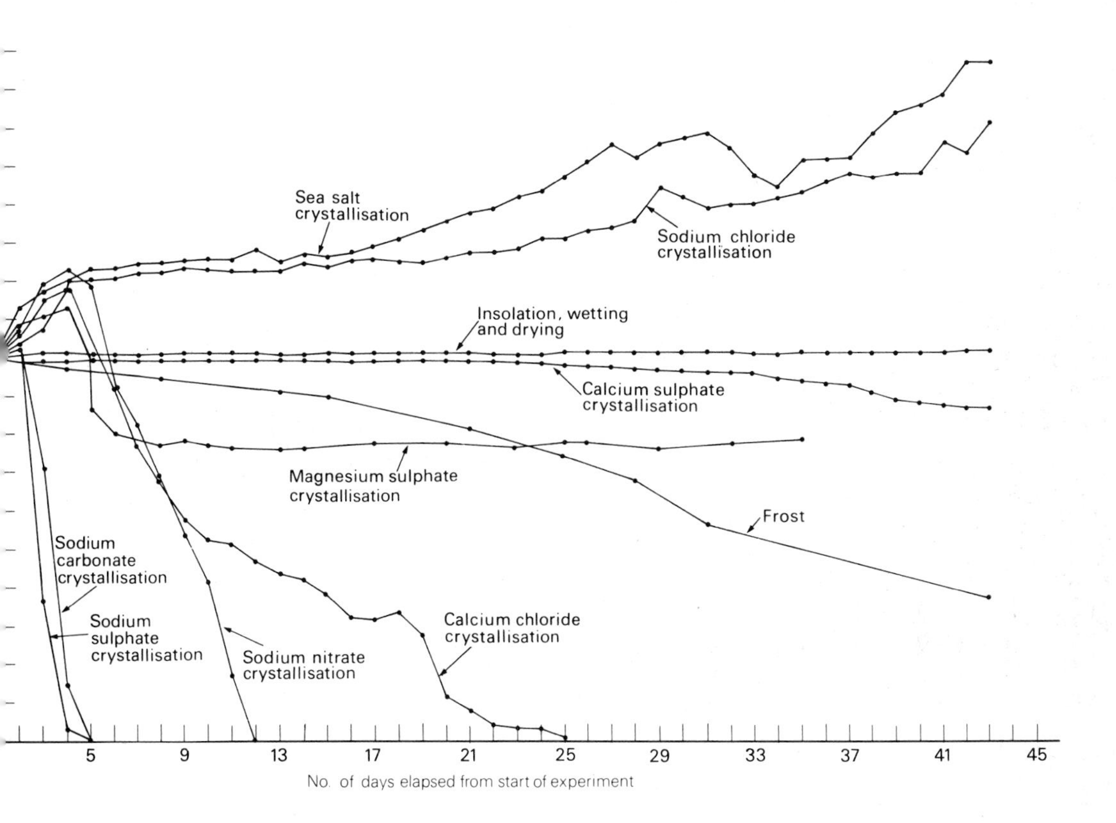

.so take place, producing skins of 'desert varnish' on rock surfaces nd leading to case-hardening, i.e. hardening of the outside of rock ırfaces. As rainfall totals and vegetation cover become higher owards the humid margins of the desert, chemical weathering will ıcrease in importance, and soils will tend as a result to become etter developed.

Even within deserts, it is often clear that rock is most severely eathered where for microclimatic or topographic reasons water is ıost freely available. Studies on the weathering of ancient Egyptian atues in the Nile Valley show this clearly. Moreover, studies of the ıoisture content within boulders in the Mojave Desert by Roth ıdicate that it is present in appreciable quantities, so that it is robably unrealistic to envisage weathering as having taken place in ıe total absence of moisture (Fig. 15).

The results of some experimental work undertaken on chalk to ssess the relative efficiencies of some different physical weathering rocesses are shown in Fig. 16. Cubes of chalk with dimensions of cm were subjected to various processes in the laboratory: insolaon (daily heating and cooling), wetting and drying, frost, and salt rystallisation with saturated solutions of different salts. Twentyour hour cycles with approximately 'natural' temperature ranges ere utilised, and the weights of all particles weighing more than .5 g were recorded daily. From this work, the relative efficiency f some forms of salt crystallisation over, for example, insolation nd wetting and drying is evident.

Aridisols

Arid environment soils–*Aridisols*–have certain well-marked features. They tend to have a low organic content because of the low *biomass* (see p. 23) of deserts, and thus are dominantly mineral soils of an immature and skeletal type. They also are little subjected to leaching so that soluble salts tend to accumulate in the profile at a depth related to the depth of moisture percolation, or to the position of the water table. Such deposition forms one of the few distinct horizons in Aridisols. It occurs nearer to the surface as precipitation diminishes (Fig. 17).

When groundwater is present near the surface, even intermittently, concentrations of salts can occur at high enough levels to be toxic to plants. Soils of this type with a saline horizon of sodium chloride ($NaCl$) in which sodium salts exceed about 2% of the mineral matter are called *Solonchaks* (white alkali soils), while those soils with sodium carbonate (Na_2CO_3) present in the upper horizons tend to be black (due to the solubility of organic matter in alkaline solutions of sodium), and are called *Solonetz* (black alkali soils).

High saline content in soils (the origins of which are described on p. 67) is second only to water availability as a factor in vegetation growth. Salt is a hazard in three main ways. First, it affects the physical structure of the soil. A high content of sodium salts, for example, leads to a dispersal or deflocculation of the soil particles or aggregates, causing a deterioration in the soil structure, giving a badly aerated, impervious soil which is inhospitable to plant growth. Secondly, there is the osmotic effect caused by a high salt content. This opposes the entry of water into roots of plants and thus increases total moisture stress. Thirdly, there is a straightforward nutritional effect, whereby toxicity or nutritional imbalance is caused: some salts in large quantities are poisonous.

Another characteristic of desert soils is their generally low clay content compared with humid-zone soils (Fig. 18). Clay content increases with increasing rainfall. The fact that arid-zone soils tend to be predominantly sandy or silty in texture means that if there is no crust development they may drain freely, sometimes too freely for satisfactory irrigation. Another characteristic of the arid-zone soils is that they tend mainly to have low organic contents, rarely exceeding 3%. As a consequence, for optimal agricultural use arid soils need nitrogen fertiliser.

17 The depth beneath the surface of the zone of calcium carbonate accumulation plotted against mean annual rainfall for United States soils.

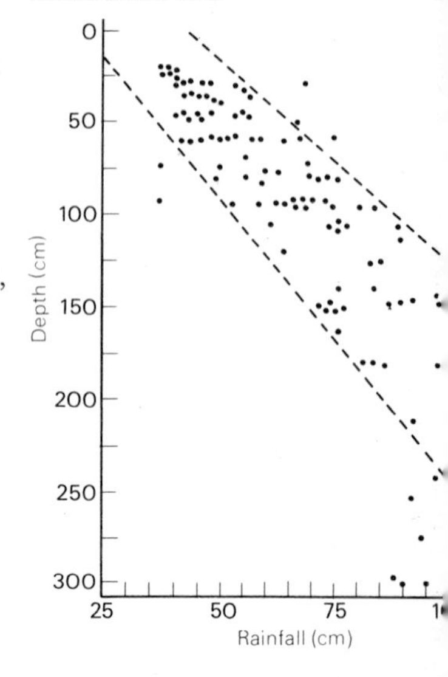

18 Clay content (colloidal material finer than 2μm) as a function of mean annual rainfall.

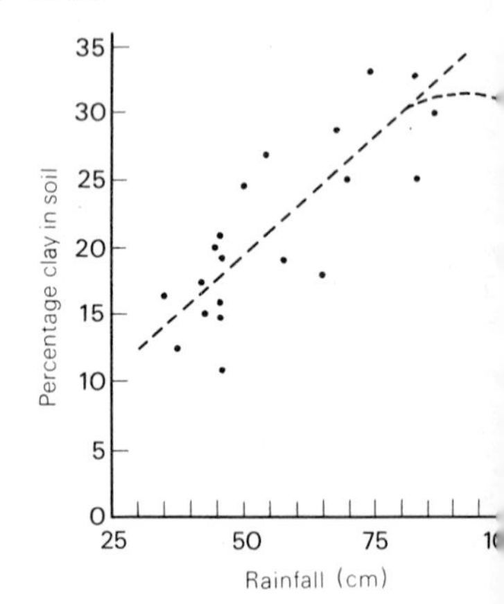

3

Desert vegetation

Water and plants

Just as the climates, landforms and surface materials of deserts are highly variable, so are the vegetation characteristics; but certain common characteristics are encountered in all deserts of the world. The most obvious one is scarcity, with the vegetation cover varying from nothing to an open stand. A closed cover is seldom attained. The second main characteristic is the seasonality of the vegetation.

Water is vital for plant growth and its scarcity in deserts is thus a major influence on desert plants' characteristics. It is vital for a variety of reasons. First the plant's protoplasm (living material) functions only in the presence of water and most tissues die if their water content falls too low. Secondly, the plant nutrients in the soil are dissolved in water and water is the medium by which they enter the plant and move from cell to cell. Thirdly, water is a raw material in the process of photosynthesis. Fourthly, water regulates the temperature of the plant by its ability to absorb heat and through the fact that water vapour lost to the atmosphere during transpiration helps to lower plant temperatures.

Biomass

A useful measure of the degree of vegetation development in an area is plant biomass, the total amount of living plant material above and below ground. Deserts have a low biomass. The wormwood and saltwood deserts of Central Asia, for example, have a biomass which is at least 100 times less than that of an equivalent area of temperate forest. In the hotter and drier deserts of the world biomass is often zero. Table 8 gives a list of biomass values for some major world vegetation types and enables a comparison to be made between deserts and other zones.

Table 8. Biomass in various major vegetation zones

Zone	Biomass (cntr/ha)*
Tropical rain forest	5000+
Broad-leaved temperate forest	3700–4100
Northern taiga spruce	1000
Savannah	700
Dry savannah	200–300
Steppe grassland	250
Dry steppes	100
Semi-shrub desert	40–45
Subtropical desert	10–15

* One centner (abbreviated cntr) = 50 kg;
One hectare (abbreviated ha) = 2.47 acres.

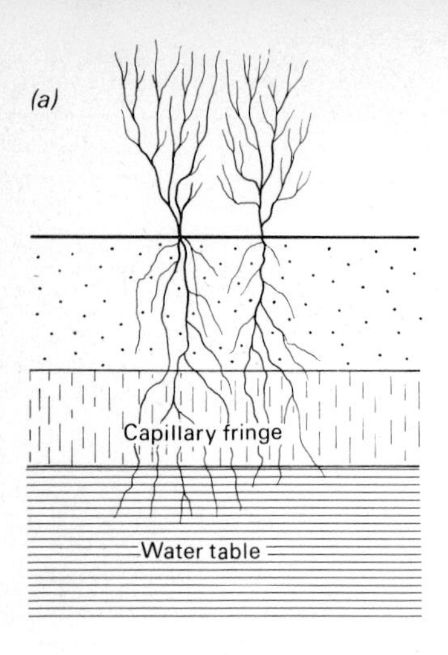

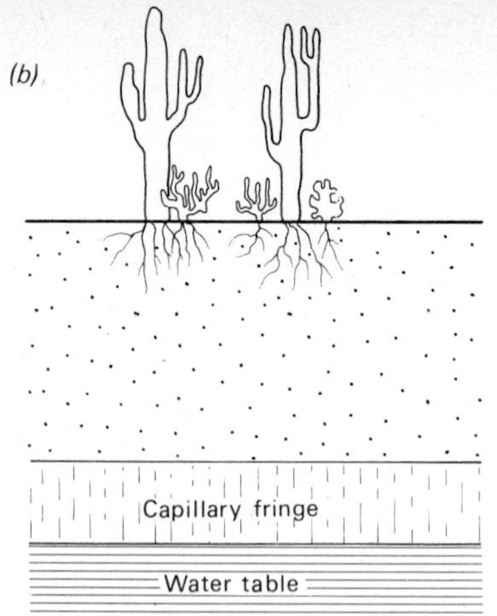

19 Diagrammatic representation of two of the major types of desert vegetation. (*a*) Phreatophytes, which derive their moisture by tapping groundwater; and (*b*) xerophytes, which are capable of resisting drought.

Classes of vegetation

There are two general classes of vegetation: perennial plants which may be succulent and are often dwarfed and woody, and annuals or ephemerals which may form a fairly dense stand after rains. The latter tend to complete their growth in a short time (6–8 weeks) and provide a grazing crop of some importance. Perennials are frequently spiny (as in the case of the Cactaceae), are leafless during the drier parts of the year and do not provide very satisfactory grazing. The cacti of the American deserts and the euphorbias of the Old World deserts are two of the main groups of succulents, though the majority of desert perennial species are non-succulent.

Most desert plants are probably best classified as *xerophytes* (Fig. 19*b*), which are plants that have the ability to resist drought and to some degree salinity. Many of them possess drought-resisting mechanisms; transpiration is reduced by means of dense, hairs covering waxy leaf surfaces, the closure of the stomata and the rolling or shedding of leaves at the beginning of the dry season. Other plants impound water in their leaves, roots and stems; these are the *succulents* (Figs. 20 and 21) and they include cacti, the prickly pear, and the euphorbias. Another way of countering drought is through a large root network relative to leaf development above ground, and it is not unusual for the roots of desert perennials to extend downwards to more than 10–15 m. Some plants are woody in type (Fig. 22), an adaptation designed to prevent collapse of the plant tissue during wilting.

Another class of desert plants is the *phreatophytes* (Fig. 19*a*). These have adapted to the environment by the development of long tap roots which penetrate into, or near to, the water table. Amongst these plants are the date palm, tamarisk and mesquite. They commonly grow near stream channels or springs, or on the margins of lakes.

These two main vegetation types, the xerophytic and phreatophytic, refer primarily to perennial species of desert vegetation which adjust to average climatic conditions, mainly by *avoidance* of drought through the mechanisms outlined above. In addition there are numerous species which comprise the *ephemeral* vegetation. These tend to *evade* drought. Given a year of favourable precipitation such plants develop vigorously and produce a large number of flowers and fruit at the end of the growth period. This replenishes the seed content of the desert soil. The seeds germinate in the next wet year and the desert blooms. These annuals sometimes possess

0 A typical group of small woody shrubs with succulent leaves growing on a rocky surface in one of the driest parts of the Namib.

1 One of the most bizarre of desert plants is Welwitschia, which occurs solely in the Namib Desert. These plants are extremely slow-growing, often being some hundreds of years old, and they exist where the precipitation is only 25 – 30 mm per annum. A camera case gives the scale.

22 This aloe, from the south-west Kalahari, shows many characteristic features of desert vegetation. It has a highly woody and tough structure, and normal leaf development is replaced by succulent spikes. It grows where the precipitation is around 100 – 150 mm per annum.

23 The rose of Jericho (*Anastatica hierochuntica*). (*a*) A plant from a watered locality; (*b*) a plant from a very dry habitat.

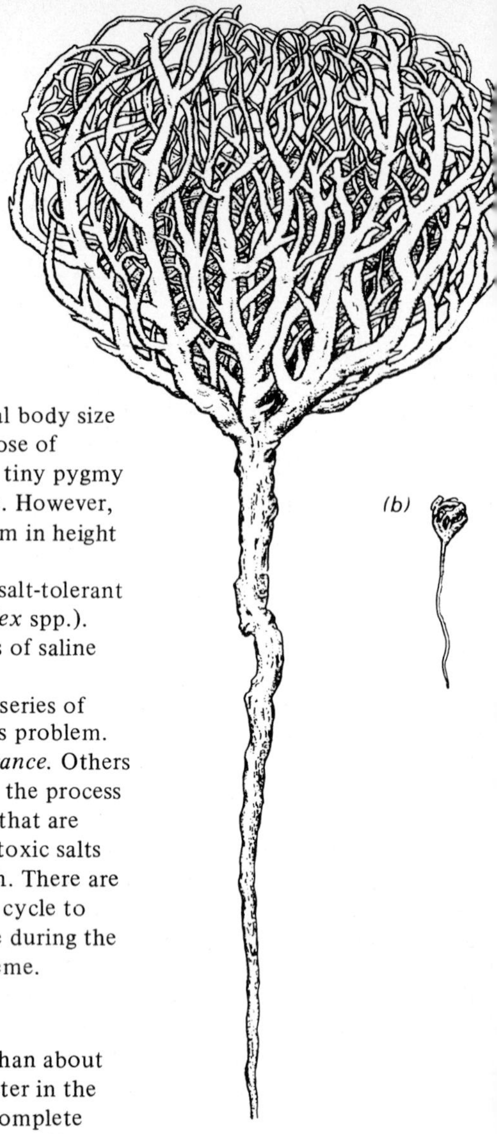

the remarkable ability of being able to regulate their total body size according to the water conditions of their habitat. The rose of Jericho, for example, illustrated in Fig. 23, tends to be a tiny pygmy a few millimetres high with few branches, leaves or fruits. However, where soil moisture is plentiful the plants reach 15–20 cm in height with tens of branches and hundreds of fruits.

One other group of desert plants is the *halophytes* or salt-tolerant plants. These include a wide variety of saltbushes (*Atriplex* spp.). They grow in areas of highly saline soils and on the edges of saline lakes.

Since salt adversely affects most vegetation growth, a series of mechanisms has evolved in desert plants to cope with this problem. Some, like the succulent *Salicornia,* have developed *tolerance.* Others have developed the ability to *avoid* salt hazard, either by the process of *exclusion,* whereby they can exclude from entry ions that are toxic in high concentrations, or else by the *excretion* of toxic salts through glandular cells when concentrations become high. There are also those plants which *evade* salt by regulating their life cycle to ensure that germination and vegetative growth take place during the rainy season, when salt hazard is less likely to be so extreme.

The nature of the plant cover

One striking characteristic of areas where rainfall is less than about 100 mm per annum is the unequal distribution of the water in the soil, and this is reflected in the plant cover. The almost complete lack of a closed cover leads to very pronounced surface run-off, so that higher ground shows relatively little infiltration of precipitation, while water penetration tends to be highest in hollows. Thus while the vegetation is 'randomly' distributed over the whole area in rain-rich regions, showing only a greater density in hollows, it changes into a more and more 'contagious' distribution in extreme arid areas (Fig. 24) so that the bulk of the area is barren and vegetation is more or less restricted to hollows.

Indeed, it is possible to identify a series of niches in arid areas each with different water relations due to differences in soil texture, topography and run-off, and these are, from driest to the wettest:

(1) Slopes with impermeable soils and high rates of run-off;
(2) flood-terraces with surface compacted soils;
(3) clay soils on level flats with high evaporation rates;
(4) sand flats easily penetrated by rain;
(5) stable sand dunes;
(6) talus (scree) slopes and stony plains with low evaporation rates;

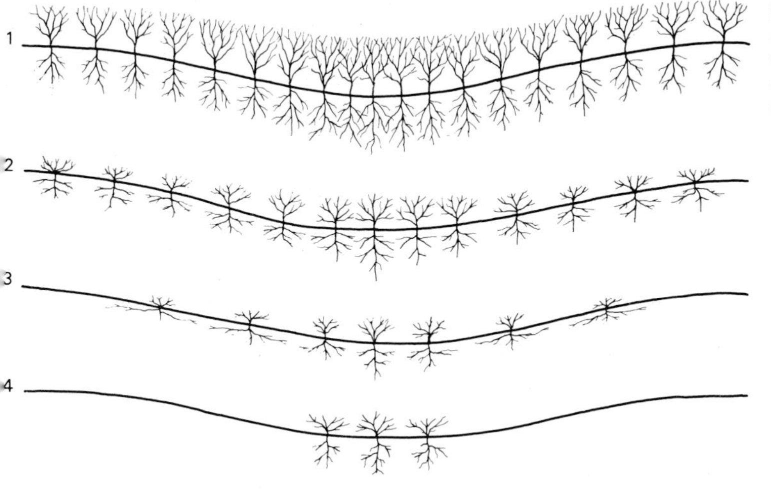

24 Schematic representation of the change from a 'randomly' distributed vegetation (1 and 2) to a 'contagiously' distributed one (3 and 4) with decreasing rainfall in extreme arid areas.

(7) fissured rocky outcrops which can hold water easily;
(8) bottoms of slopes with inflow of seepage;
(9) erosion channels carrying water temporarily and storing water in the underlying soil;
(10) dry valleys containing a continuously flowing stream of water at no great depth below the surface.

Aerial photographs of semi-arid areas, particularly where rainfall is between 100 and 400 mm per annum frequently reveal curious regular patterns produced by differences in vegetation distribution and density. These patterns found, for example, in Somalia, resemble the black markings of a tiger skin, and are thus termed *brousse tigrée* by the French. They tend to occur either as arcs or stripes, and may sometimes be of considerable size, with a spacing of 50-200 m. The patterns are formed either by special grass communities or by dense stands of acacia, and do not occur on flat areas or on very steep slopes. Their origin is still not clear, and more than one mechanism may be involved in their formation. However, in many cases it seems probable that they develop as a result of sheetfloods depositing arcs of sediment and water favourable to plant growth on low-angle slopes. Thus this *brousse tigrée* is good evidence of the importance of local soil, moisture and geomorphic factors in determining the nature of desert vegetation distribution.

4

Surface forms and processes: wind

Wind erosion and deflation

During the late nineteenth and early twentieth centuries various geomorphologists, especially the Germans Passarge and Walther, postulated that wind erosion was a predominant cause of desert erosion and that in particular the presence of isolated hills (*inselbergs*) and surrounding plains could be considered as peculiarly desertic phenomena caused by severe undercutting of rock masses. Wind erosion was also held to be responsible for certain bizarre minor landforms such as shaped stones called ventifacts, yardangs (shaped rock masses with rounded upwind faces and long, pointed down-wind projections; see Fig. 25) and zeugen (yardangs composed of rocks of unequal resistance and of a complex form associated with the presence of a resistant cap rock). Some depressions were also held to be formed by wind erosion of bedrock (Fig. 26).

Later there was a shift of ideas against the power of wind erosion to mould whole desert landscapes, and until recently wind was relegated to a relatively insignificant position as an erosive process and the role of desert rainstorms was stressed instead. More recently, the examination of aerial and satellite photographs has led to something of a return to ideas of the power of wind erosion, in that these photographs show some massive grooves and troughs in desert areas. For example, on the west and south-east sides of the Tibesti Massif in the Sahara such features (Fig. 27), which are between 0.5 km and 1 km wide and are spaced from 500 m to 2000 m apart, dominate the topography over some 90,000 km^2. The regularity of their forms and their alignment with the curving path of the prevailing north-easterly winds lend very strong support to the idea that they are wind erosion features.

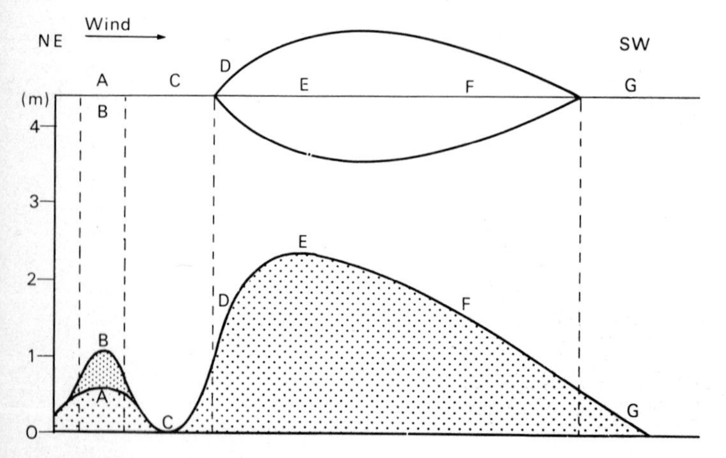

25 Yardang morphology from the Sahara. A, Upstanding sandstone aureole (see Glossary); B, crest of wind-blown sand; C, depressed sandstone aureole; D, front; E, head; F, reverse; G, train.

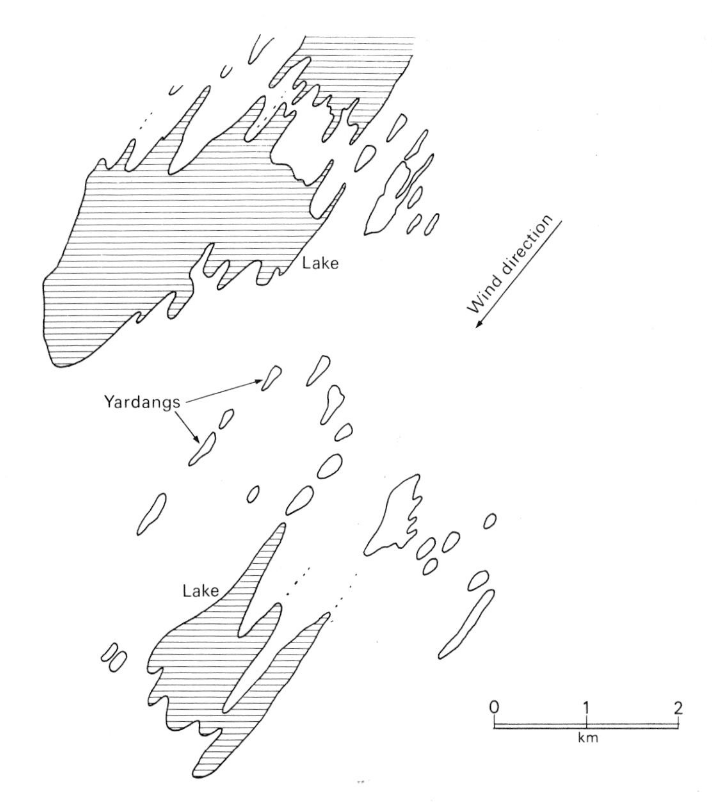

26 Lakes and yardangs produced by erosion and deflation by the wind at Ounianga Kebir, Ennedi.

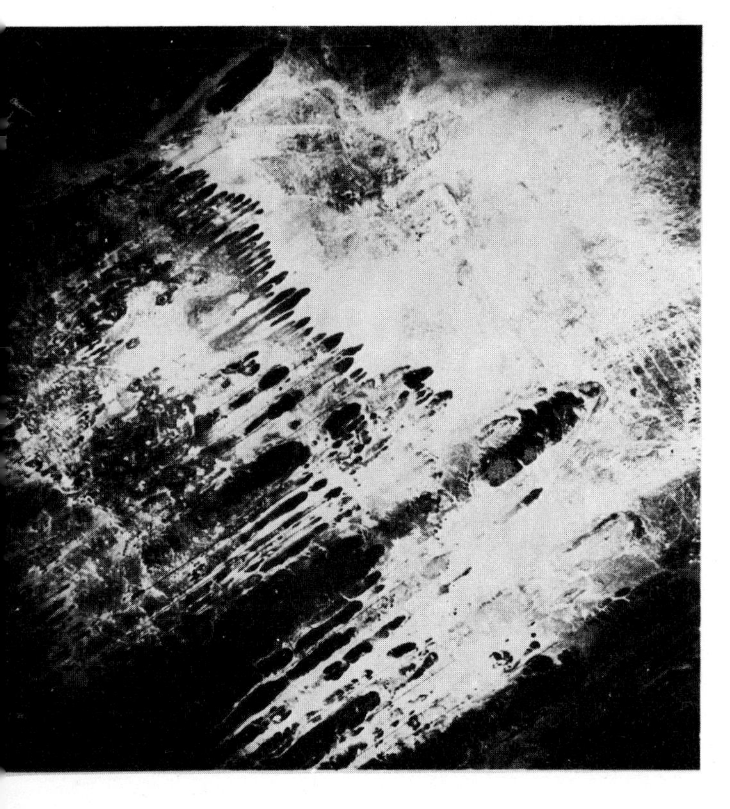

27 Although the power of wind erosion in deserts has come to be doubted in many areas in recent years, the study of satellite photographs and aerial photographs (such as this one from the Sahara) do show up remarkable areas of grooved terrain which parallel the direction of the trade winds.

28 The exposed roots of this mesquite tree in the Thar Desert of India indicate the extent of the soil erosion that has taken place in the area as a result of population pressure and overgrazing by herds of goats. Many trees in the area a also cut down for firewood, thereby increasing the erosion hazard.

In addition, the recognition that many closed depressions have an aeolian origin has led to a re-appraisal of the importance of wind. As Peel wrote in 1966: 'the volume of material excavated from the Qattara I estimate roughly at some 800 cubic miles, and, if this and similar depressions are truly the work of wind, it seems an over-statement to write it off as of little account as an agent of land-sculpture'.

In addition to the role of wind erosion in forming such landforms, deflation is also an insidious and widespread cause of soil removal (Fig. 28). Dust storms resulting from deflation may block roads, fill ditches and canals, bury crops, disperse insects and weeds, remove seeds, expose plant roots, and deplete the soil of its humus, in addition to the simple physical process of bulk soil removal.

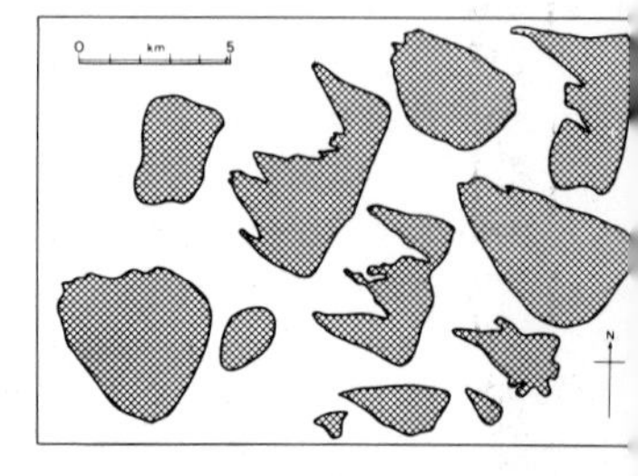

29 The shape of wind-moulded pans from the south-west Kalahari Desert, South Africa. The formative wind came from the north west.

The deflation process, which, as we have already seen, is one process which may lead to desert pavement formation, is an effective one in arid lands for a variety of reasons. First, the soils, because of low silt, clay, organic matter and moisture contents, frequently cannot resist wind action. Second, lack of vegetation means that there are large areas over which wind action can operate, wind velocity is not checked, and there is little plant residue to protect the surface and to add organic material to it. Moreover, recent German studies suggest that some desert whirlwinds attain a speed of 126–158 km/h (hurricane force).

The main factors involved in the process of deflation, then, are the erodibility of the soil, soil surface texture (in general a rough surface is more effective in reducing wind velocity than a smooth one), local climatic factors involving wind velocity and precipitation, the length of exposed ground along the prevailing wind direction, and the quantity of vegetation cover.

Desert depressions

Depressions are an important feature of all deserts and are variously called playas, pans, flats, dayas, deflation basins, and so forth. They have various forms and origins.

The extensive playas of the south-western United States are fossil lake basins of the Pleistocene pluvials, originally produced by faulting in the Basin and Range province. Water still remains in some of them, but there are over 100 dry basins in this area.

However, many of the depressions are not purely structural and

30 A daya formed on the desert limestone surface of the Qatar peninsula, probably as a result of localised solution. These shallow depressions sometimes provide a source of freshwater, which can be tapped by wells of the type shown in the photograph.

are the result of other geomorphic processes, especially deflation by the wind. In the Kalahari there are probably over 9,000 depressions, and some of them show orientation with the wind (Fig. 29), characteristic clam shapes, and lunette dunes on their lee sides (see Fig. 33). Other depressions occur along lines of old river systems and represent relicts of channels which have become blocked by dunes, though, once formed, both wind action and solution may have played a role in their enlargement. The dayas of desert limestone surfaces (Fig. 30) may be formed by solution, as in Fig. 31*b*.

Another mechanism that may well have produced desert depressions is animal activity. Before the decimation of the herds by hunters in recent centuries, semi-arid regions were richly inhabited by game animals. The animals would tend to concentrate around any hollow where water might accumulate, and by pawing and removal of mud on their bodies would tend to enlarge the initial depression to produce, for example, the *buffalo-wallows* of the Great Plains. Thus desert depressions, like many other desert landforms, illustrate the principle of equifinality, whereby different processes can lead to the development of broadly comparable forms (Fig. 31).

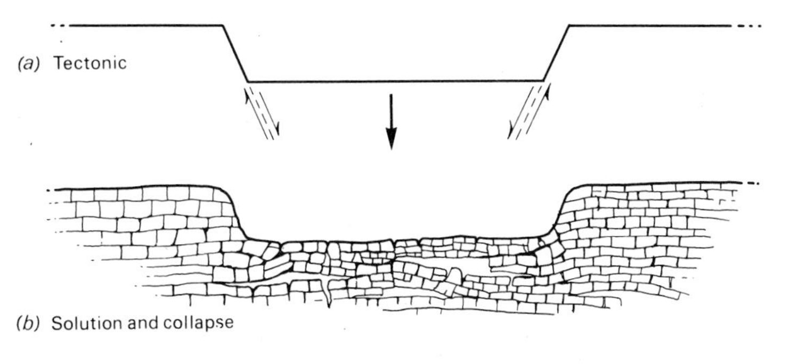

31 Desert depressions form in a number of different ways. Some, (*a*), may be tectonic, others, (*b*), may be formed by solution of limestone and other soluble rocks, and others, (*c*), may form by deflation.

32 The Umm al-Samim (pronounced as 'Sameem'), a vast playa, receives much of the areic drainage from the Oman mountains. The polygons of the surface are formed by the desiccation of the salts and clays as the water evaporates.

In whatever way the depressions have been formed, they are distinct environments, and tend to have a surface of either fine debris, such as clay and silt, or materials produced by evaporation of groundwater (evaporites) which include sodium chloride and sodium carbonate (trona) deposits. They vary in the degree to which they are subject to flooding, and they vary in size from a few square metres to the 9300 km^2 of Lake Eyre. The depressions are important components of the desert environment in that they are sites of mineral wealth (especially chlorides, sulphates, carbonates, nitrates and borates), they are attractive for military and high speed vehicular activity, and they may be foci for human settlement and animal herds (as are the Etoscha and Makarikari pans of the Kalahari). Frequently the surfaces of desert depressions are covered by large polygonal patterns produced by the desiccation of the surface salt or clay crusts (Fig. 32).

Sand deserts

Only about one-third to one-quarter of the world's deserts are covered by aeolian sand, so its role in deserts should not be exaggerated. Indeed, in the American deserts sand dunes occupy less than 1% of the surface area. The importance of dunes lies in their role in water supply (see p. 62) and in the fact that they are one of the most regular and striking of landforms. Great *ergs* or fields of regularly formed sand dunes are found in no other parts of the world, except in a somewhat attenuated form around some of the great ice caps. They may attain remarkable heights, some dunes

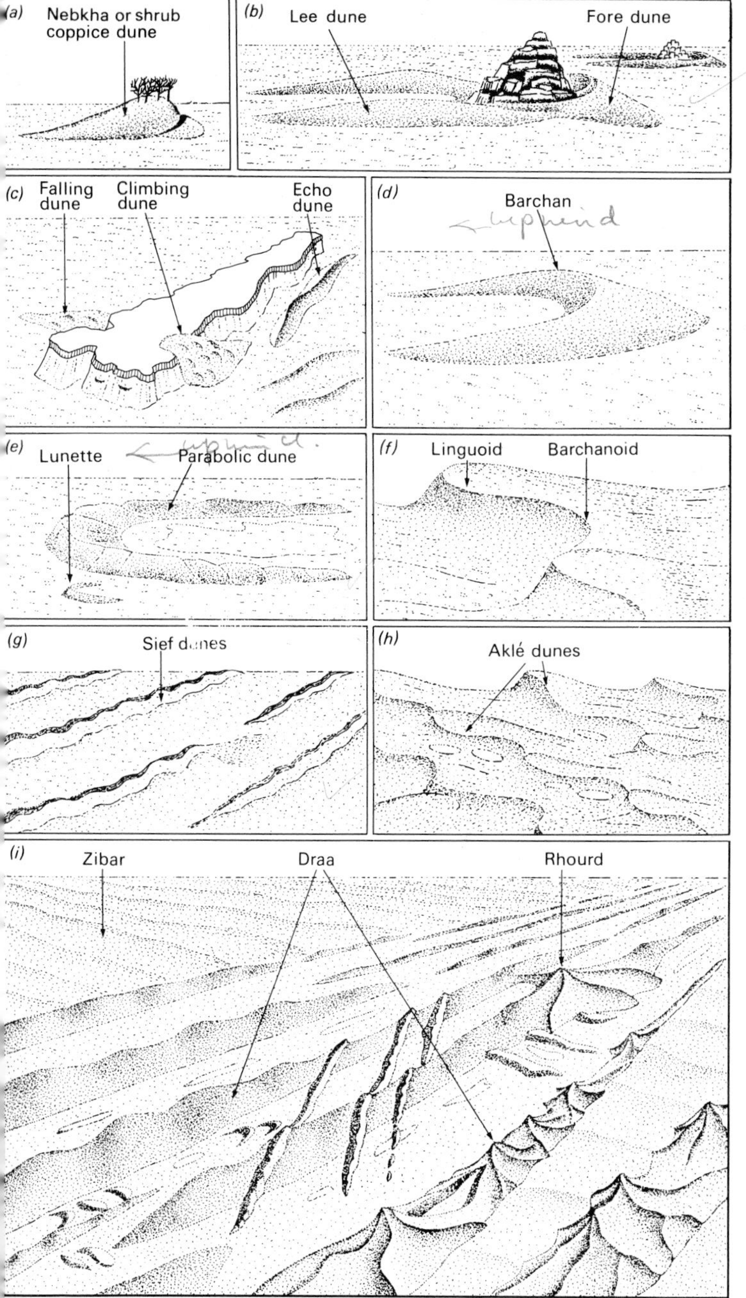

33 A selection of the main types of desert dune forms.

in South West Africa (Namibia), the Lut of Iran and the Empty Quarter of Arabia, exceeding 200–250 m.

The attractions of the ergs have been well described by Bagnold: 'Instead of finding chaos and disorder, the observer never fails to be amazed at a simplicity of form, an exactitude of repetition, and a geometric order unknown in nature on a scale larger than that of crystalline structure. In places vast accumulations of sand weighing millions of tons move inexorably, in regular formation, over the

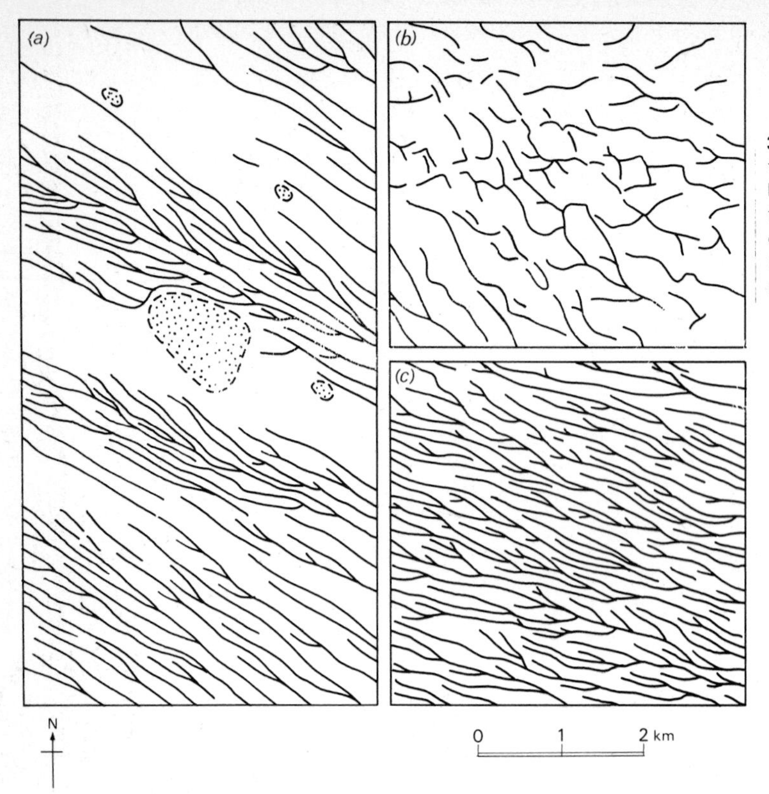

34 In the south-west Kalahari dunes tend to have Y-junctions rather than a simple linear form. This morphology occurs in three main patterns: (*a*) clustered dendritic, (*b*), reticulate, (*c*) dendritic. Form (*c*) is geometrically similar to many river systems.

surface of the country, growing, retaining their shape, even breeding, in a manner which, by its grotesque imitation of life, is vaguely disturbing to an imaginative mind.'

The variety of sand dune types is considerable (Fig. 33), though many textbooks give the impression that there are only two main types – the crescentic *barchan* and the longitudinal – but the old distinction between transverse and longitudinal forms does not make allowance for the numerous oblique forms that are encountered during aerial photographic studies (Fig. 34).

One class of dune is that formed by the interference of an obstacle, such as a hill, with wind and sand movement (Fig. 40). Sand may be deposited both to the lee and the windward of such obstacles, and the many names for these dunes include lee dunes, fore dunes, falling dunes, echo dunes, and wrap-around dunes. Fig. 35 shows a wrap-around dune from the Thar Desert in India. Other topographic dunes develop in the lee of desert depressions. They are crescentic in shape and are called *lunettes.*

Most dunes, however, develop without topographic interference. The crescentic barchan is easily distinguishable and has its horns pointing upwind. Barchans are in fact rare, and it has been calculated that only 0.01% of the sand in active dunes in the world's deserts is to be found in them. Barchans form where there is a limited sand supply and where winds are unidirectional throughout the year. Another crescentic form, but one which has its horns facing downwind, is the *parabolic* or *linguoid* dune. Sometimes these two types of crescentic dune occur together to form sinuous ridges transverse to the main wind direction, and to these the name of *aklé* is given. Aklé dunes are much more common than barchans; they are one of the simplest types of network dune, and they seem to require relatively unidirectional winds and a large sand supply.

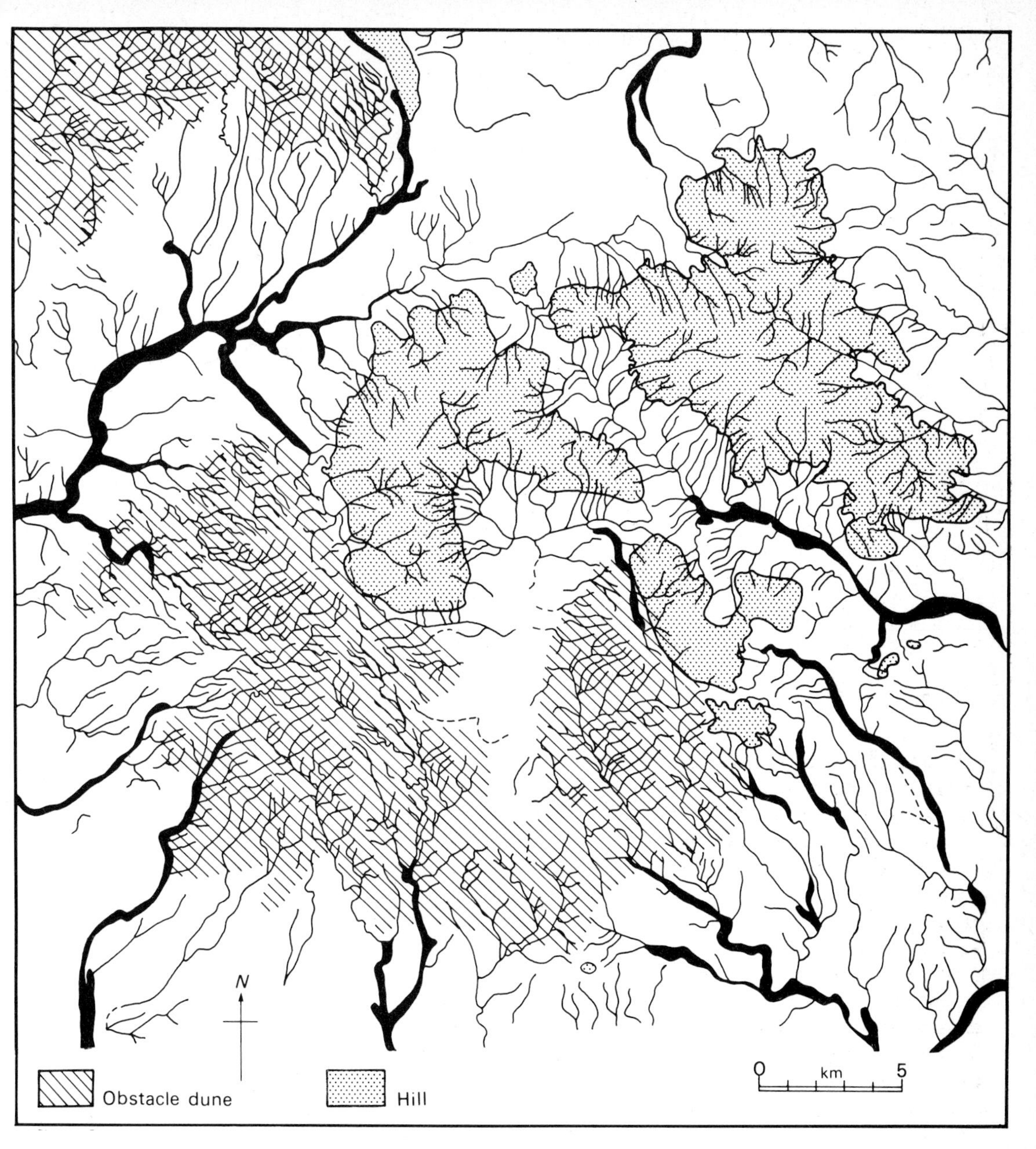

35 A large topographic or obstacle dune formed by sand being piled up against a large hill in the Thar Desert, India, by the powerful south-west monsoon.

In some deserts almost straight dunes occur in regular patterns, and these are called longitudinal dunes. They develop nearly parallel to the dominant wind direction, and cover large areas of the Empty Quarter, the western Sahara, the Libyan desert and the Simpson desert of Australia. Their forms in some areas may be complicated by the presence of Y-junctions with angles of 30–50° to the main flow (Fig. 34). More complex patterns may involve the creation of nodes where two or more dune trends come together. Such nodes may give rise to large star- or dome-shaped sand mountains called *rhourds* (Fig. 36).

Although maps or aerial photographs of ergs appear at first sight to give very complex patterns, in reality these are usually both repetitious and regular. There tends to be a hierarchy of sizes. This is illustrated in Table 9. Ripples (between 2 cm and 2 m apart) are widespread and occur on the surface of dunes, which are between

36 Desert sand dunes take many forms. These star-shaped dunes, often called rhourds, occur in the Sahara. The aerial photograph illustrates two sizes of phenomena: large sand masses of the draa category, and the smaller features of the dune category which have developed on them.

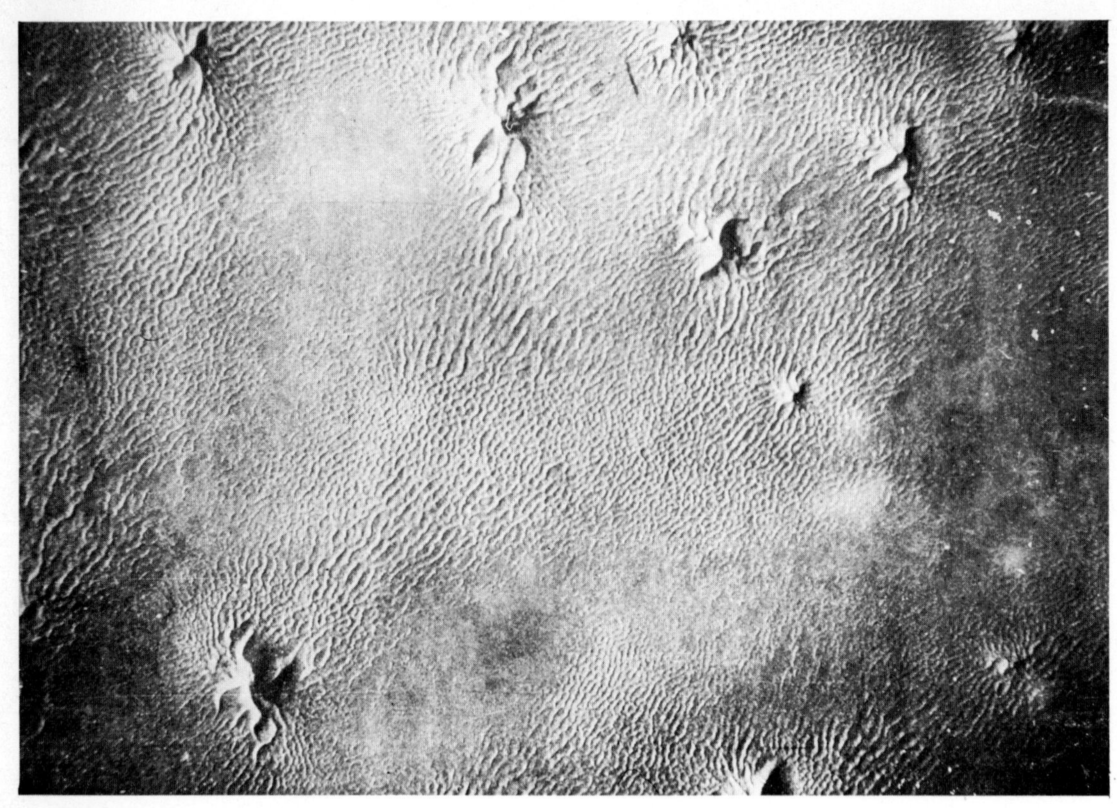

20 and 300 m apart. These dunes often occur on larger features (between 1 and 3 km apart) called *draas*, which form a sort of plinth. In any one area there is a tendency, given a constancy of wind speed and grain size, that these three distinct classes of form will occur, with few features of intermediate size.

Even at the very greatest scale of investigation, on a continental scale, dunes show a remarkable regularity of pattern. Thus in Australia (Fig. 2), North America, the Sahara and the Kalahari one has a *wheelround* of dunes in an anticlockwise direction related to dominant continental wind patterns. Another interesting but inadequately understood feature of the wheelrounds is that the dunes of which the whirls are composed have a consistent asymmetry of cross-profile, with the steeper dune flank towards the outer margin of the wheelround.

In some situations it is apparent how one dune type can grade into or develop from another. Thus Bagnold in a classic model showed that the crescentic barchan can, by moving into an area with a different wind regime or sand supply, become deformed into a longitudinal *sief* dune. Conversely, in the Thar Desert in Pakistan, Verstappen has shown that the other type of individual transverse dune, the *parabolic* or U-dune can gradually be elongated until a blowout occurs so that two longitudinal dunes are developed (Fig. 37).

Bagnold's classic model is illustrated in Fig. 38. An initial barchan (*a*), formed by a prevailing wind (*x*) becomes subject to another wind from a different direction (*y*). This tends to lead to the elonga-

ble 9. Hierarchical classification of dunes

der	Name	Wavelength (spacing)	Amplitude (height)
rst	Draa	300–5500 m	20–450 m
cond	Dunes	3–600 m	0.1–100 m
ird	Aerodynamic ripples	0.15–2.5 m	0.2–5 cm
urth	Impact ripples	0.5–2000 cm	0.01–100 cm

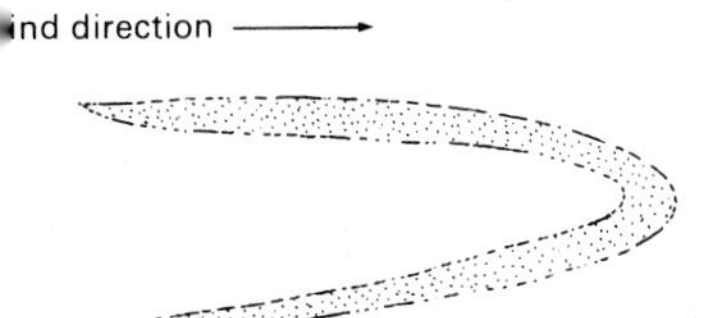

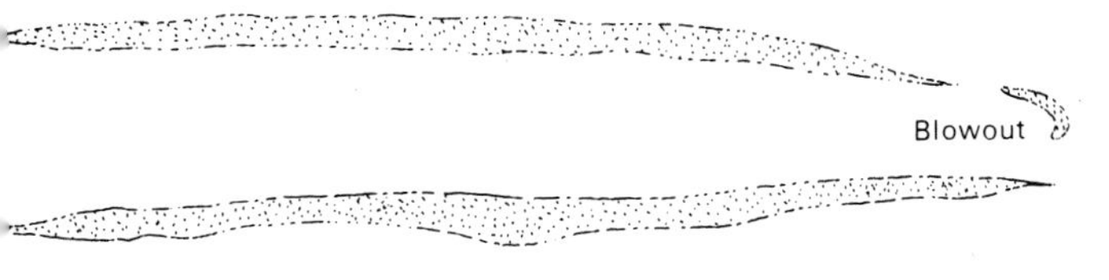

37 Some longitudinal sief dunes may develop from a parabolic dune. The original parabolic (*a*), gradually becomes elongated (*b*), until a blowout occurs (*c*), creating two parallel dunes. Compare this model with Bagnold's (Fig. 38).

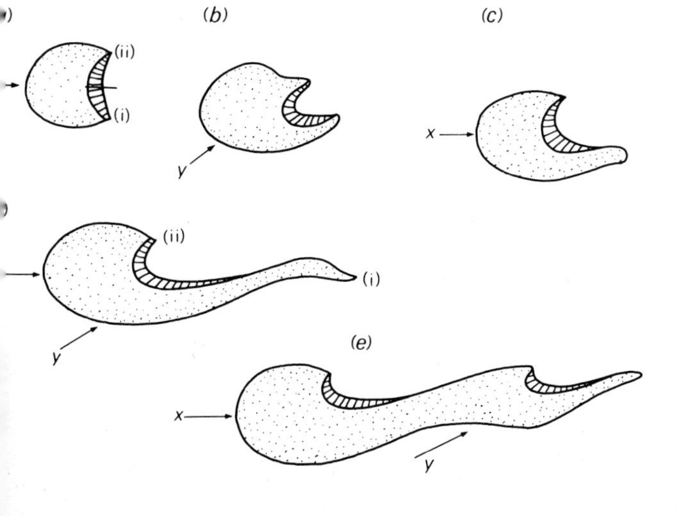

38 The Bagnold model of sief dune development from the crescentic barchan. For explanation see text.

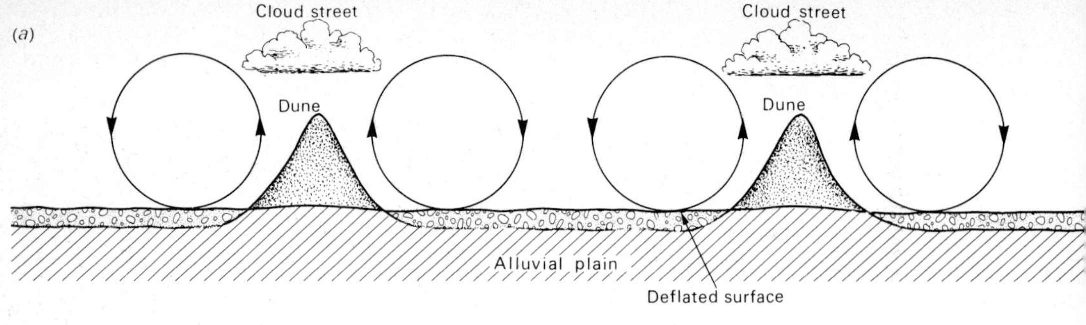

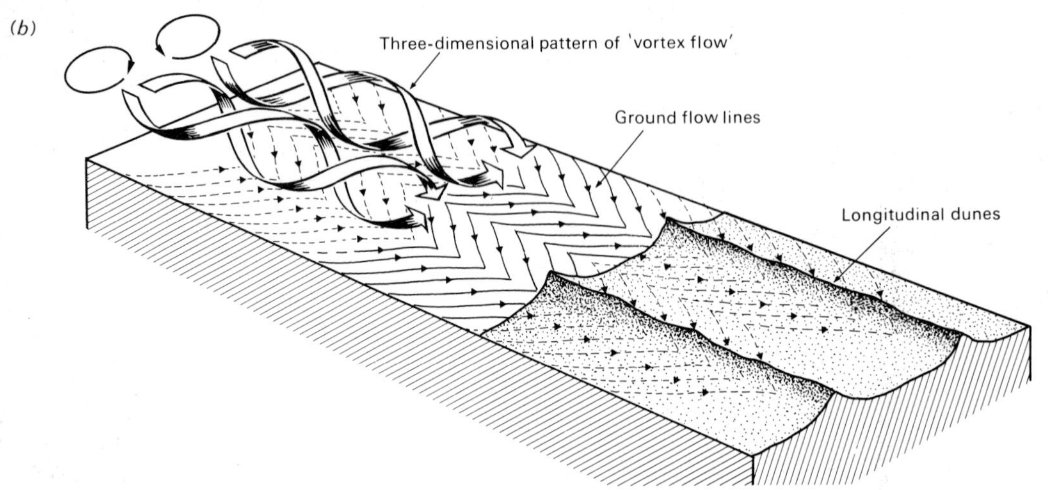

39 (*a*) A two-dimensional illustration of roll vortices and associated dunes, deflational plains and cloud streets. (*b*) A three-dimensional illustration of the type of vortices in the trade winds which could produce regularly spaced and parallel longitudinal dunes.

tion of horn (*i*) at the expense of horn (*ii*), and in time a sief dune may form.

One possible aerodynamic explanation for the long parallel sief ridges observed in many of the world's ergs involves the development of longitudinal helical-roll vortices. These features are illustrated in Fig. 39. That such vortices exist is testified to by the presence of long lines of shallow cumulus clouds called *cloud streets.* The cloud streets, when they have been observed, appear to have the same order of spacing as sief chains.

Sometimes desert dunes become cemented by percolating water (Fig. 40), and this enables one to study the characteristic cross-bedding of aeolian deposits (Fig. 41).

40 One major form of desert dune is the obstacle dune, formed by the banking up of sand against a suitable barrier. Here a cemented calcareous dune sand, locally called miliolite, derived from the warm coasts of the Arabian Sea in N.W. India, has accumulated against Junagadh Hill and is now quarried for building stone.

41 (below) Aeolian sand is characteristically deposited with the cross-bedded structure shown in this photograph of a semi-cemented dune (aeolianite) in Bahrain.

5

Surface forms and processes: water

The work of water in deserts: run-off

The nature of the run-off pattern in deserts is often one of short-lived high-peaked floods produced by a combination of occasional torrential showers and favourable ground surface conditions. The desiccated surface of the desert is, except in the case of sandy tracts, relatively impermeable to water and little rain seeps into the ground; much of the water is held at the surface rather than infiltrating. The lack of vegetation, the lack of an absorbent organic layer in the desert soils, and the presence of hardpans and soil crusts all contribute to rapid run-off, and thereby contribute to the effectiveness of the rain in the process of erosion and transportation of debris. Thus is spite of their low precipitation, deserts are areas where fluvial action may be important.

Desert stream channels or valleys are often called *wadis* (Fig. 42). This is not a very precisely defined term and means little more than a stream course which is normally dry but sometimes subjected to large flows of water and sediment. The term has been applied both to deep, flat-bottomed, rock-cut gorges, such as one might find in a mountainous desert like Sinai, to shallow, braided and frequently shifting 'washes' developed on long-angle slopes like pediments and fans. Some wadis will show both characteristics in their courses.

The regime that one might expect to find in a small desert river is illustrated in Fig. 43 where the run-off patterns for two wadis in the Central Sahara are depicted. High flows with discharges of 20–60 m^3/s are separated over a short time-span by periods of almost no flow at all.

In the Sahara there is an average of one flood a year in the semi-arid parts, but in the more arid Tademait only six or seven floods in ten years. In areas of almost total aridity the wadis may stay ten years without carrying water. It is also common for any one stream only to carry discharge along a portion of its course, and floods often peter out because of seepage and evaporation. Saharan floods seldom succeed in flowing more than 300 km. When they do flow, such rivers often carry a considerable sediment load, and they also create problems for those trying to travel (Fig. 44).

Sediment yield in deserts

By measuring the suspended load in rivers or by recording the quantity of accumulated material behind reservoir dams it is possible to obtain a measure of the sediment yield for any given area in m^3/km^2 per year. Clearly, if there is no run-off, then the sediment yield value will be zero. However, as we have already discussed, run-off may be significant in many areas, and the absence of vegetation means that surface soil and weathering debris is moved with some ease. In areas with enough precipitation to give quite high degrees of

2 The lack of vegetation and soil cover n steep rock slopes, such as these in the inai Desert, encourages rapid run-off fter rain so that in very dry areas there may be large wadis of impressive power. Palm trees and acacias make use of the greater moisture on the wadi margins.

43 The discharge pattern (hydrograph) of two central Saharan wadis in one storm in November 1958. (*a*) is a wadi at Issakarassene (Hoggar), with an area of 140 km², and (*b*) is a wadi at Timesdelsine with a catchment area of 1500 km². Flow is both short-lived and irregular.

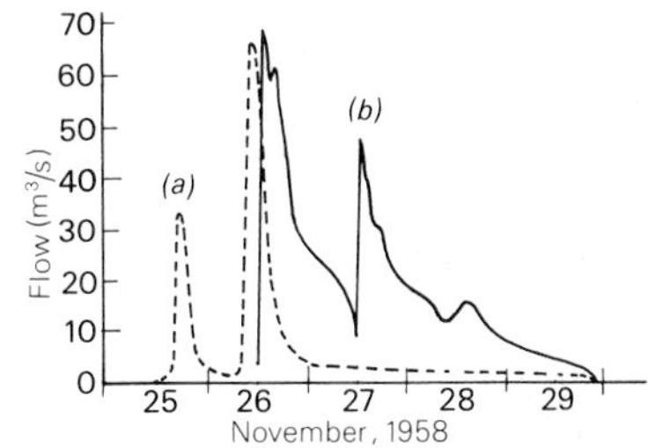

44 Desert stream courses may suddenly fill with water after rainfalls. This produces the so-called flash flood, which caught one of the authors unawares in the Sumayil Gap, a line of weakness through the Oman mountains etched by wadi systems in the junction of igneous rock (low relief on the left) and the limestones and dolomites which form the Jabal Akhdhar massif (on the right).

un-off but not enough precipitation to give a dense and erosion-educing vegetation cover, one might expect to encounter quite high ates of sediment yield. Work in America by Langbein and Schumm, hough it is based on data which are in some respects imperfect, has hown that not only is this the case, but that in semi-arid areas ediment yield is at a maximum (Fig. 45). In desert areas, as precipi-ation rises above zero, sediment yields increase at a very fast rate ntil they are damped down by the presence of a dense grass cover, vhich serves to protect the soil from rain splash impact and other rosive processes.

Gully development

n many desert margin or semi-arid areas this observed tendency owards high rates of sediment removal is reflected in the rapid levelopment of ephemeral gullies called *arroyos* in the United States nd *dongas* in southern Africa. They develop especially rapidly in more easily eroded materials, such as alluvium, and can cause great

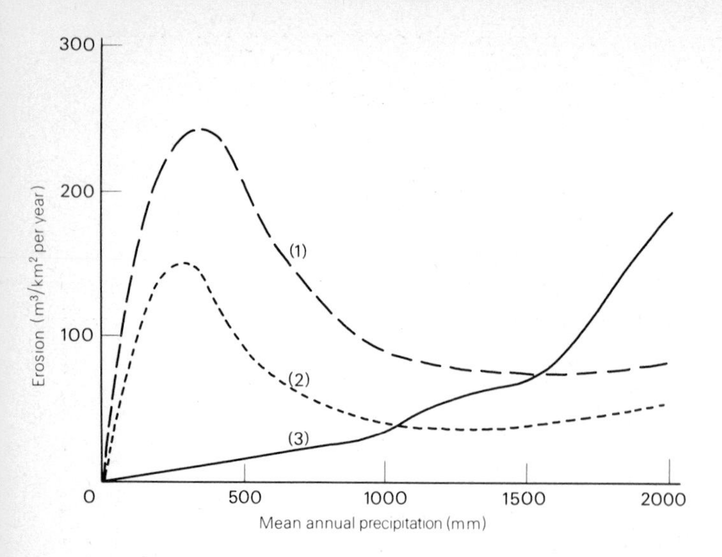

45 (1) Represents rates of erosion in high-relief situations in relation to mean annual precipitation; (2) represents rates of erosion in lower-relief situations; and (3) represents rates of limestone solution. Rates of erosion appear to be very high under semi-arid conditions, whereas rates of solution only become high under conditions of substantial rainfall.

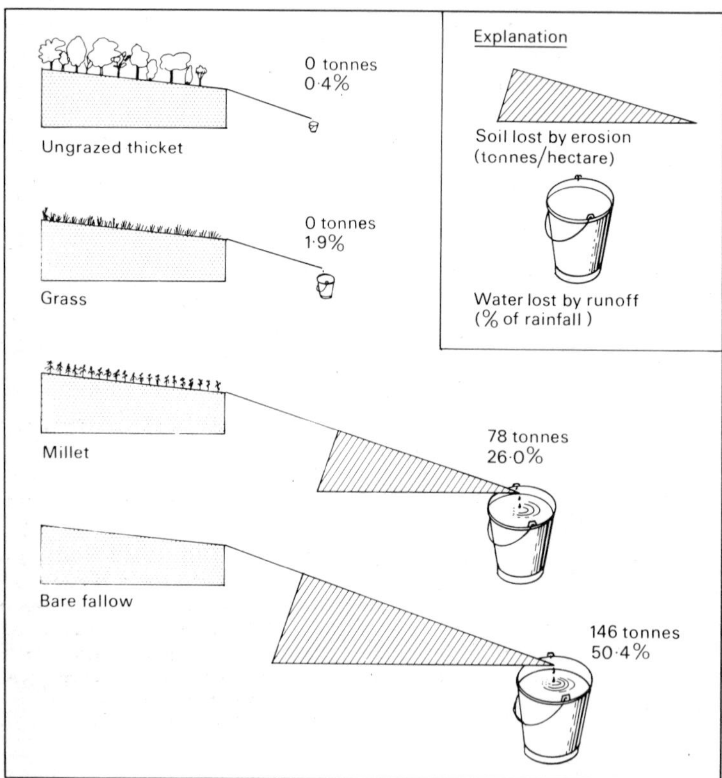

46 Results of soil erosion tests on ground with different vegetation covers in a semi-arid part of Tanzania. The grass-covered plots had little loss of soil and water, whereas rates of soil loss and run-off from bare fallow ground were very high.

damage by disrupting roads, entrenching and dividing fields, and lowering groundwater levels.

The reason why these dangerous features develop as suddenly as they sometimes do, is an instructive example of the principle of equifinality which we have already referred to in the context of desert pavement and desert depressions. Basically, however, they result either from an increase in erosive stress or from a decrease in resistance of the materials (e.g. the alluvium) on which the erosion is operating. Changes that take place in the stress and resistance may either be natural – for example, an increasing frequency of heavy rainstorms may cause increased erosion – or they may be caused by

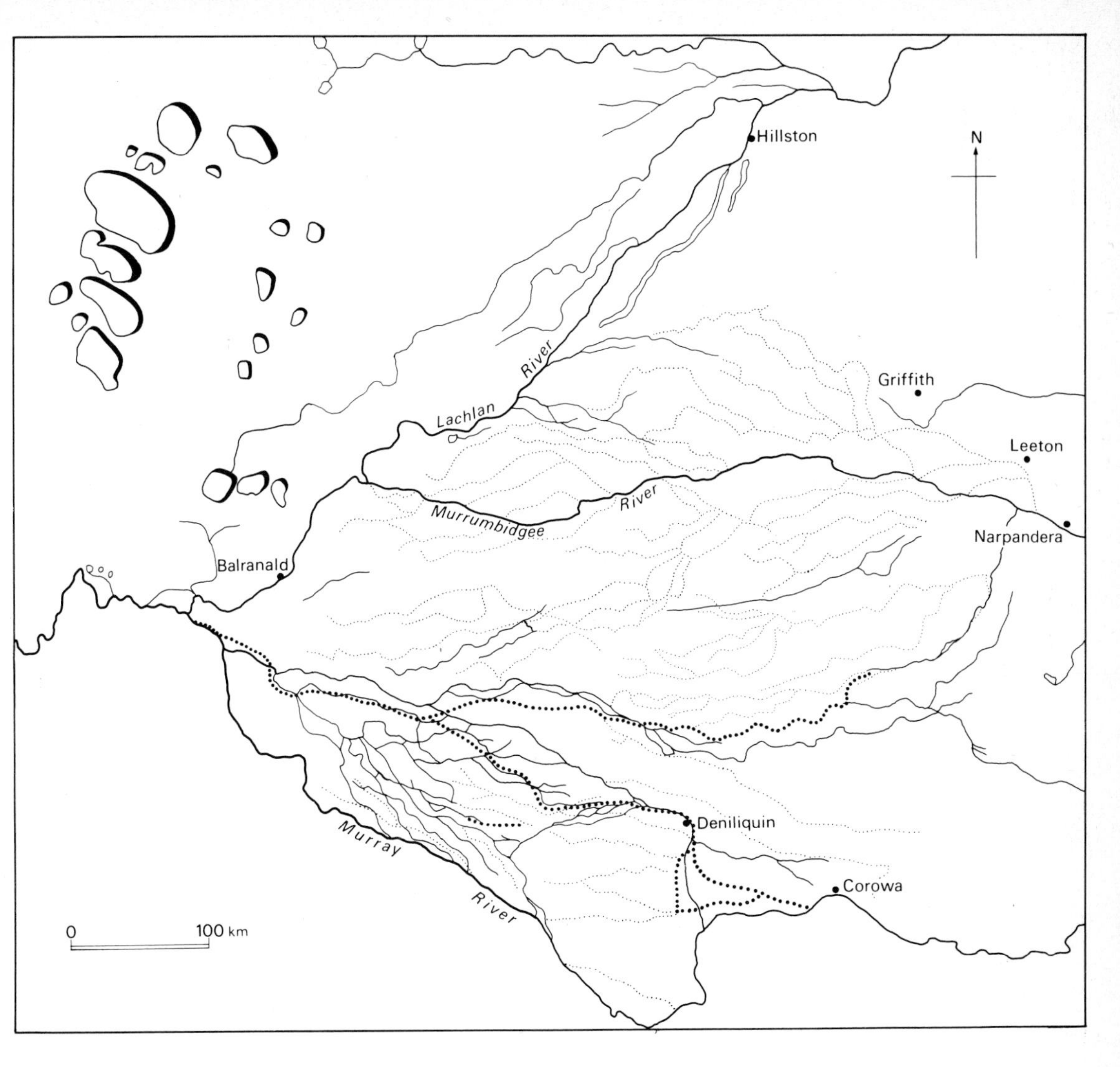

47 Part of the Murray River Basin in New South Wales, with present streams shown by continuous lines, prior streams by dotted lines (or wider dotted areas for a prior stream complex). Lunettes, which are crescentic dunes that develop in the lee of desert depressions (see p. 34) are shown by solid black areas on lake shores.

man, who, for instance, could decrease the resistance of alluvium to erosion by removing its grass cover, by deforestation, or by trampling of the surface by domestic animals. Man may also increase erosive stress locally by constructing roads and other structures which may channel water and focus it onto a particular point on the alluvium, thereby causing it to be cut away.

Connected with this process of gully development is the no less insidious process of soil erosion. Especially on the desert margins, disturbance or removal of the vegetation cover by man can markedly increase rates of sediment yield. The scale of this problem is well illustrated in Fig. 46 which portrays the results of a recent survey by a Swedish–Tanzanian team in semi-arid parts of Tanzania. Soil loss was virtually nil under ungrazed thicket or grass, rose to 79 tonnes/hectare under millet, and to almost 146 tonnes/hectare under bare fallow. There was also a comparable increase in the amount of water lost by run-off. Water infiltration was far less under bare fallow conditions.

Desert drainage systems

The examination of the present and past climatic conditions of deserts, in Chapter 1, has already shown that nearly all present-day

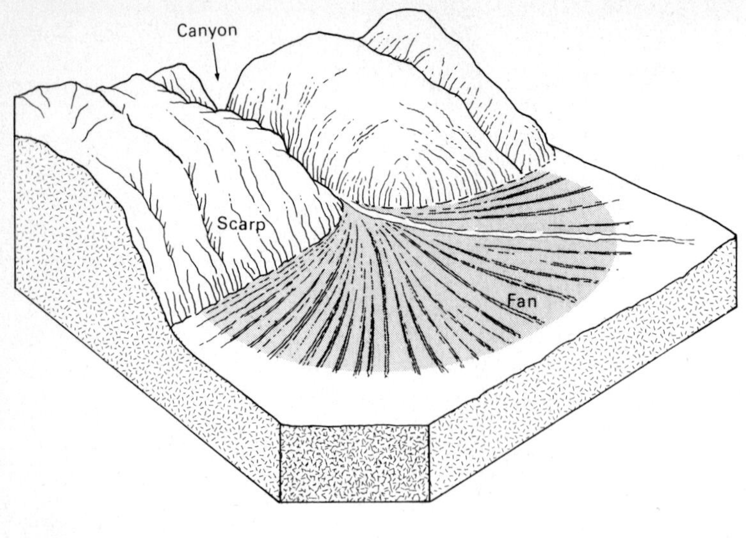

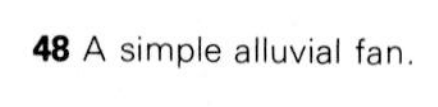
48 A simple alluvial fan.

deserts receive precipitation which is adequate to cause run-off, albeit locally and for a short time, and that in the past they were subjected to more humid phases called pluvials. Thus although deserts are by definition relatively dry areas, they still carry drainage systems, and any differences between them and more 'normal' areas are essentially a matter of degree. Much of the drainage, because of high evaporation or because of infiltration into alluvial or aeolian materials, does not reach the sea, and because of the existence of closed basins produced by deflation and other processes of the type illustrated in Fig. 31, much of the drainage may be *endoreic* (flowing towards the centre, i.e. centripetal). It also needs to be remembered that some rivers in deserts, like the Nile, the Indus, and the Tigris–Euphrates, have their sources outside the true desert realm, and thus they may impose alien hydrological and geomorphological conditions upon the desert scene. Such rivers are sometimes called *allogenic* or *exogenous* rivers.

Some desert river systems differ in degree from those in other zones of the world's surface in that they show more disorganisation and less co-ordination. This is illustrated by a study of the river system of south-eastern central Australia, where there are innumerable dry sandy river beds of an ephemeral nature which have, over millennia, wandered extensively over their plains built up by deposition of sediment. The various tributaries of the Lachland, Murrumbidgee and Murray systems form the complex pattern illustrated in Fig. 47, rather than the more conventional dendritic or tree-like pattern of normal systems.

The study of such systems is important in that the old or 'prior' stream systems constitute collecting channels for underground water, and the intricate pattern of ancient fluvial deposits influences both underground water development and irrigation and drainage schemes in the Riverina area.

Alluvial fans

A major component of the desert drainage system is the alluvial fan (Figs. 48 and 49), which can be defined as a cone-like deposit of alluvial material laid down beyond the limits of valleys or canyons because of a change in the nature of flow when a wadi leaves the confines of its mountain channel. Such fans are not confined to hot desert regions – they do, for example, occur widely in periglacial

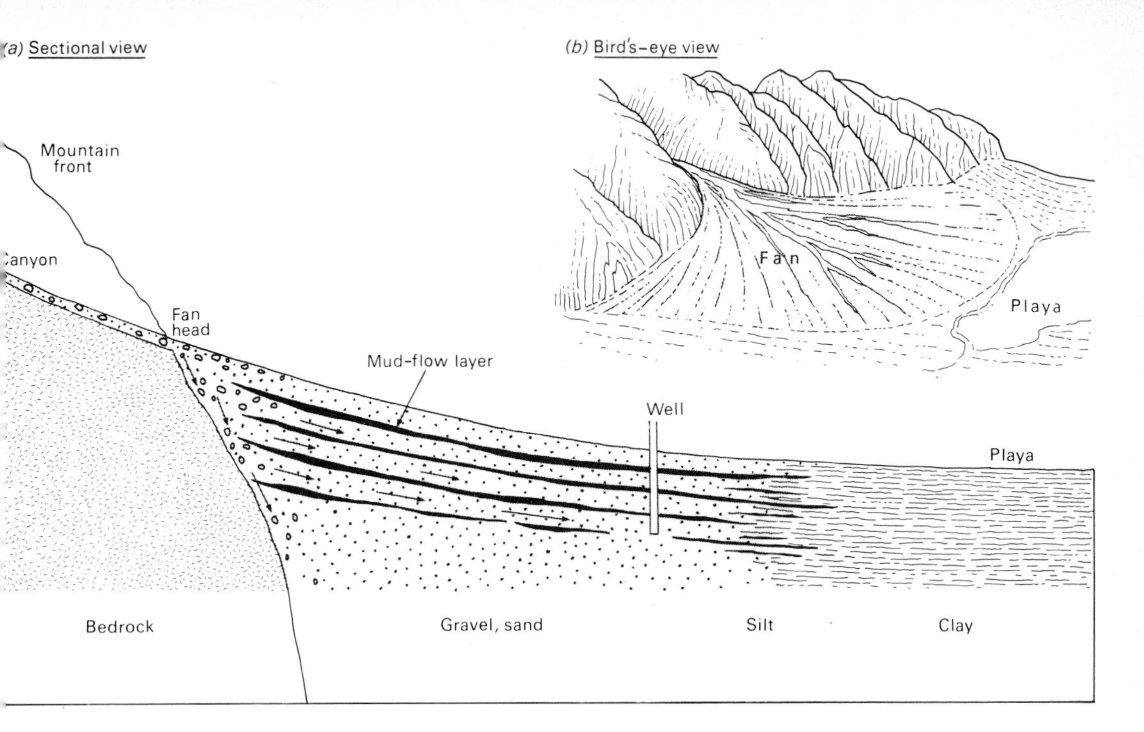

49 Idealised cross-section of a complex alluvial fan showing the movement of groundwater from fan head through aquifers of gravel and sand.

reas – but various factors lead to their particularly fine development in hot deserts. First, a lack of vegetation means that channels an shift. Secondly, occasional heavy thunderstorms evacuate highly ediment-charged masses of waters from the mountains. Thirdly, arid-one weathering processes may produce considerable quantities of oarse debris. Another necessary requirement for their formation is hat there should be a sharp juxtaposition of mountain and lowland, nd for this some tectonic activity may be required as in the Basin nd Range province of the western United States.

The size of the fans, which may be considerable, is largely determined by the basin area of the contributing stream and the degree of esistance of the rocks in the basin. Fans derived from easily eroded ocks are larger than those derived from resistant rocks. The types of low which deposit the material comprising the fans vary from imple streamflows to highly viscous debris flows. Because of the ariability of flow regime and the variable nature of the material omprising the fans, they are subject to rapid changes in character, hannels shifting laterally over a wide area, and channels alternately *ncising* (cutting) and *aggrading* (filling) themselves. This presents problems for the establishment of transport links and buildings on an surfaces. On the other hand they may be locations with useful roundwater resources (see p. 65).

Desert slope systems

Slope profiles in arid landscapes are very dramatic in their impact ince they tend to be more angular than in other areas, and they are ittle obscured by vegetation (Fig. 50).

The forms of desert slopes can be analysed in terms of Wood's dealised slope profile. This has four components: an upper convexity (waxing slope), a cliff (free face), a straight segment constant slope), and a basal concavity (pediment). In particular there is often a very abrupt *break of slope* between the last two components. These four components are commonly all present in

50 Representation of a characteristic semi-arid slope form after King, Wood and others. The pediment slope is essentially a feature developed in bedrock – it is not a depositional surface. The constant slope is a rock-cut slope thinly covered in debris. The upper slope of the plateau surface, being an old surface, frequently possesses a capping of ancient soils and weathering crusts.

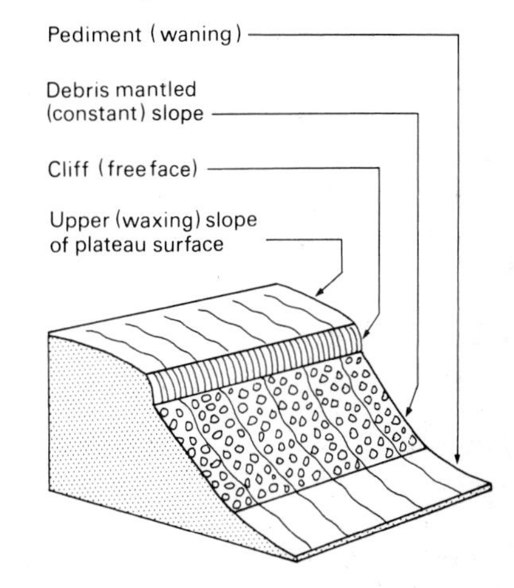

desert profiles, though sometimes either the cliff face or the straight slope may be absent. However, it is unusual for both of these steep-slope components to be absent, whereas this is more usual in humid temperate areas. The basal concavities and associated plains may cover extensive areas, and from them may rise isolated residual outliers called *buttes,* which possess one or both of the steep-slope components.

The form of the straight slope and the basal concavity may need some explanation. The straight slope looks basically like a scree or talus slope, but it differs in that the debris on it usually forms only a veneer above a bedrock slope. It is not so much a slope of accumulation (like a true scree or talus) as a slope where there is an equilibrium between material being supplied from the cliff face and material being removed downslope. It is a slope of transportation and any fine material on this slope is removed by wash processes. Coarse material remains until it is reduced by the processes of mechanical and chemical weathering into something fine enough to be removed by wash processes.

The pediment problem

Although alluvial fans are a common phenomenon in some deserts between the mountain front and the alluvial basin, another landform type which also occurs in this marginal position is a relatively narrow bedrock plain called a pediment, forming part of the basal concavity. The literature on this is both lengthy and controversial, particularly as regards pediment origin. The form was first described in the American south-west, but has now been identified widely in Africa and in India.

In essence, pediments consist of a gently sloping surface, generally at an angle of between 1 and 7 (occasionally up to 11) degrees, which is only lightly dissected, has few lines of concentrated drainage, and which is composed of bare rock or a thin veneer of detritus overlying bedrock. Such plains are commonly faintly concave in their longitudinal sections, but straight and convex sections have been recorded. One marked feature of such pediments, which is not so often found in more humid landscapes, is that there is a sharp break of slope, often corresponding to a change in debris size, between the mountain front and the pediment itself (Fig. 51).

An early explanation of pediments was that they were formed by sheets of water of high velocity (sheetfloods) which pared down the surface to give a relatively undissected, gently sloping surface. Such sheetfloods have from time to time been described, but they are probably not very common. However, a striking argument that can be used against this theory is that whatever the effects of rare sheetfloods may be, they cannot create the pediment surface, for a relatively smooth, undissected gentle slope is a prerequisite for their establishment. A second hypothesis sees pediments as having been shaped by *lateral planation.* That is, it is envisaged that streams flowing across a mountain front would swing from side to side and gradually erode the surface. However, it is difficult to explain pediments that directly abut the mountain front away from such swinging channels. As the American geomorphologist Lustig has put it: 'The hypothesis virtually requires that streams emerge from a

51 Pediments and pediment passes in the Sacaton Mountains, Arizona.

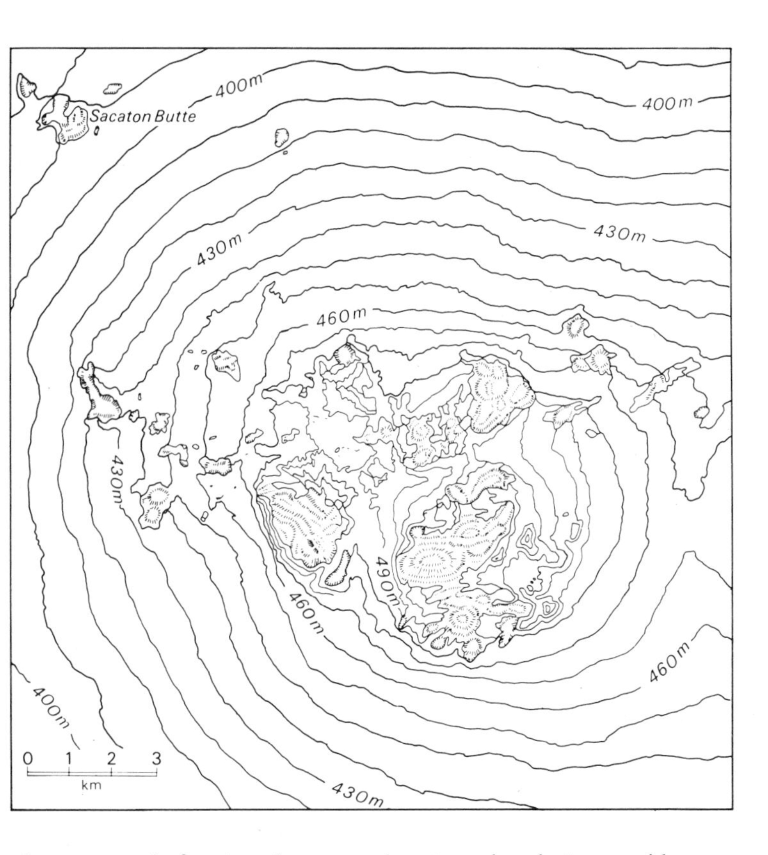

given mountain front, and, on occasion, turn sharply to one side or the other to "trim back" the mountain front in interfluvial areas. Such stream paths, nearly perpendicular to a sloping surface, would defy the laws of gravity and have not been observed except in those areas where drastic tilting has occurred'. A third hypothesis involves the role of surface and subsurface weathering. Subsurface weathering is likely to be accentuated at the junction between the mountain front and the plain because of the natural concentration of water there through percolation. This will lead to preferential weathering at a break of slope, and the products of the weathering will be removed by sheetflow, wind and other processes, so that the mountain front retreats backwards and the pediment becomes larger. When pediments coalesce and cover large areas they are called *pediplains*. Lester King, a South African geologist, believes that such surfaces dominate the African landscape.

It is probable, as with desert depressions, stone pavements, and other desert landforms, that pediments may result from more than one process, and one should not perhaps expect to find that one hypothesis can account for all examples.

Inselbergs

Rising above and behind pediments are often isolated, steep-sided hills called *inselbergs.* Many of these features develop as a result of escarpment recession (Fig. 52*a*) and pediment development of the type outlined above. They occur in a wide variety of rock types. The famous example of Ayers Rock in the Northern Territory of

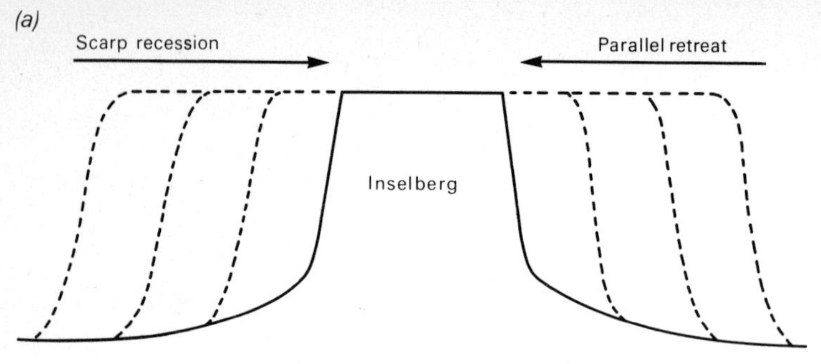

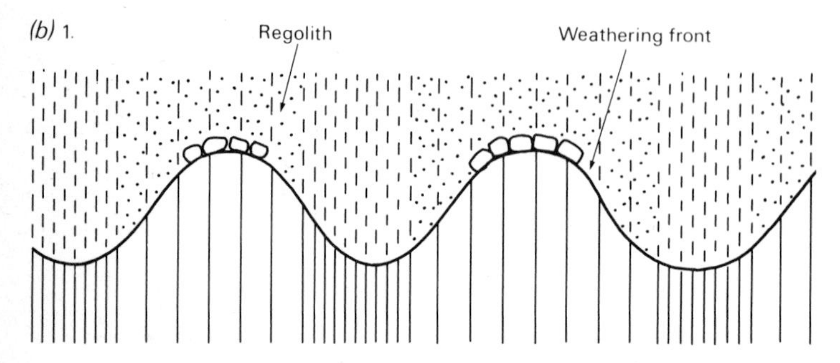

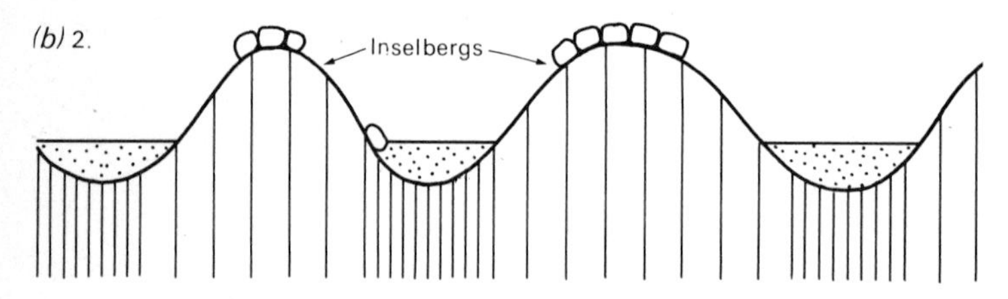

52 The two main models of inselberg formation; (*a*) by parallel slope retreat, and (*b*) by the stripping of deeply weathered material.

Australia, for instance, is formed of sandstone. Domed inselbergs, on the other hand, are especially well-developed in granite, the structure of which renders it especially prone to inselberg development (Fig. 53).

However, not all inselbergs occur in deserts, nor do they all necessarily result from the process suggested by King (Fig. 52*a*). An alternative hypothesis, which may well apply to inselbergs on the southern margins of the Sahara, is that which sees them as developing as a result of a two-stage process rather akin to that proposed for tors by D. Linton. The first stage (Fig. 52*b*1) involves rotting of bedrock, especially granites, to a considerable depth, probably under warm and moderately humid conditions. Weathering is particularly effective and reaches to a greater depth in situations where the granite is closely jointed. Areas of massive (widely spaced) jointing are relatively less affected. In the second stage (see Fig. 52*b*2), the weathering products (regolith) are removed by erosion (perhaps brought on by a change in climate or of base-level) and so the uneven *weathering front* of stage 1 is exposed, the massively jointed areas being exhumed to give inselbergs, and the closely jointed areas giving basins of low relief.

53 An oversteepened small inselberg rising above a rock-cut pediment in the Thar Desert, India. Granite frequently produces such rounded shapes in arid areas.

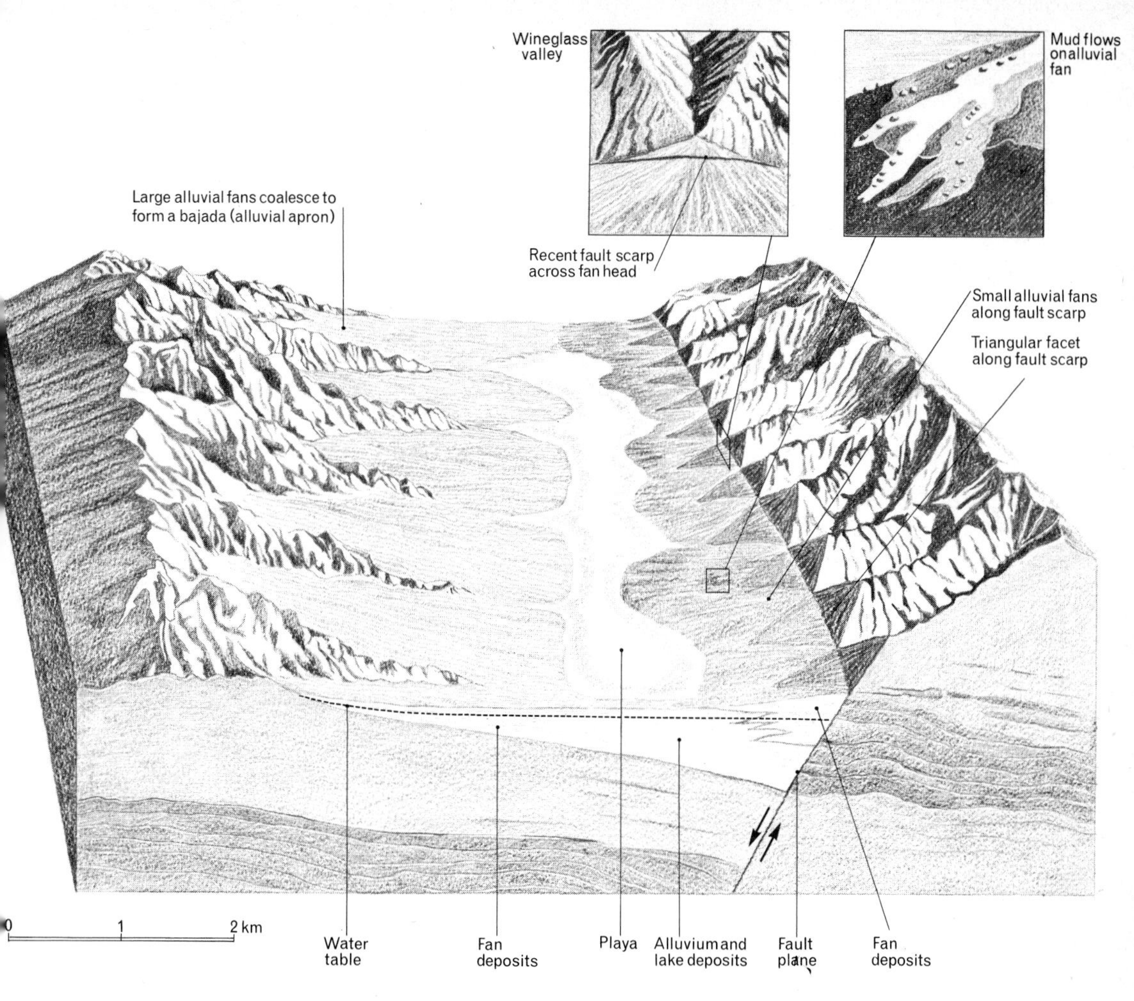

54 A generalised impression of a desert landscape in the basin and range situation. Notice the importance of fans, playas and faults.

There is reason to believe that both the hypotheses may be applicable, according to local circumstances, for inselbergs take on many different forms and vary in size from the small feature shown in Fig. 53 to massive hills over 600 m high.

The desert landscape: the basin and range situation

In any particular desert the individual landforms discussed so far, such as pediments, alluvial fans and inselbergs, tend to make up an assemblage which gives a distinct landscape. One such landscape is illustrated in Fig. 54. In the south-west United States, for example, there are alternations of basins and ranges. The basins are areas that have dropped with respect to the neighbouring ranges (uplands) as a result of earth movements. The line along which the displacement has taken place is called a fault.

Water collects in the basins, in which through-flowing water seldom occurs, to give a playa lake containing evaporites (see p. 18). In a humid area the basins would fill with water and eventually overflow to produce outlet channels. On the sides of the basins are the alluvial fans, reflecting rapid run-off from the bare uplands. When the streams reach the low-angle slope of the basin they lose much of their velocity and thus the coarse sediment settles out. The fans on the left of Fig. 54 are steadily drowning intervening spurs on the down-tilting block and sedimentation is causing local base level to rise. By contrast, on the right-hand side the fans are smaller and

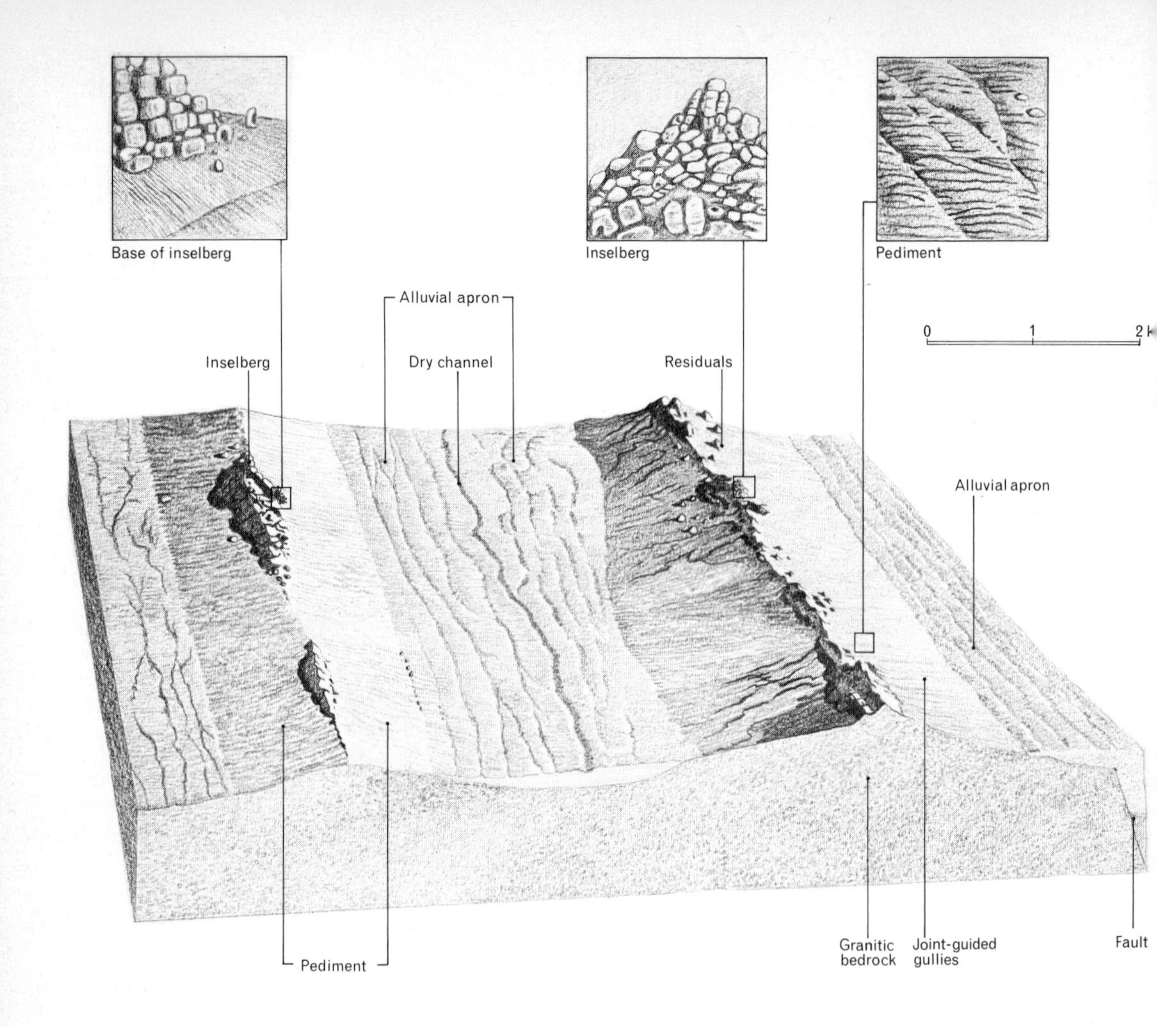

55 A generalised impression of a desert landscape in an area of crystalline rocks, demonstrating the importance of inselbergs and pediments.

steeper, and the fault scarp of the up-tilting block is clearly marked by a series of triangular facets. Differential movement causes the valleys to 'hang'. Forms are generally angular on account of a limited soil cover and the limited nature of chemical weathering. Debris tends to be angular and coarse.

A desert landscape: erosion of crystalline rocks

A second landscape assemblage to show the interrelationships of some desert landforms and processes is shown in Fig. 55. Here the landscape being eroded is composed of crystalline rocks (such as granite) and earth movements are minimal.

The landscape is dominated by long smooth ramp-like forms (composed of an alluvial apron merging upwards into a rock-cut surface – the pediment) which have steep, bouldery slopes projecting above them (the inselberg). The inselbergs rise abruptly from the pediment. *Creep* and *solifluction* have done little to reduce slope angles and slope evolution has been predominantly through back wearing. The bouldery slopes result from the tendency of granitic rocks to have many vertical joints, which are weathered by miscellaneous processes to give boulders. The lower portions of these slopes may be zones of some moisture accumulation so that weathering occurs more potently, producing oversteepening, undercutting and tafoni (see p. 19) formation. Products of weathering are transported across the pediments to accumulate in topographic lows

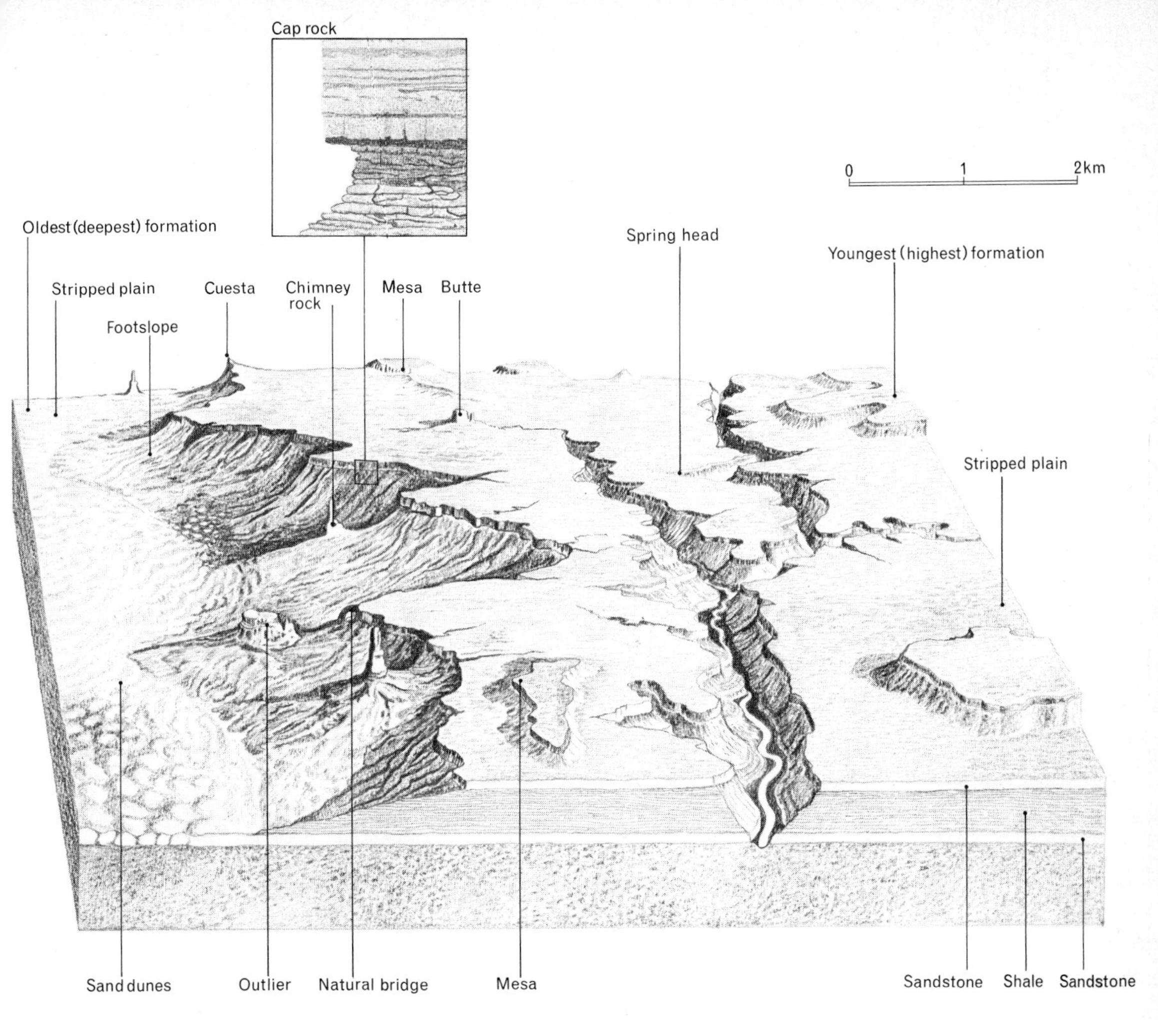

56 A generalised impression of a desert landscape developed in an area of sedimentary rocks.

where they may be re-sorted to give sand dunes.

This type of landscape is well displayed in the Aravalli Hills in the dry zone of north-west India.

A desert landscape: erosion of sedimentary strata

A final example of a landscape assemblage to show the interrelationships of some desert landforms and processes is shown in Fig. 56. The landscape in this case is developed through the erosion of sedimentary rocks with a low degree of dip.

Basically the landscape is dominated by an escarpment or *cuesta* which wears backwards in time in a down-dip direction. The top of the cliffs tends to be developed into a resistant cap rock, and cliff retreat takes place through the undermining of less resistant underlying beds. This process may be accentuated by sporadic spring sapping which creates large cavernous alcoves, some of which have provided shelter for primitive peoples, such as the Pueblo Indians of the south-west United States.

Where less resistant beds have been stripped off the bedrock there may either be a rather featureless plateau, dotted with a few deflational or solutional hollows, or a plateau above which rise a few erosional remnants called *mesas* (flat-topped) or *buttes* (remnants which are too small to preserve a flat top). Forms resulting from wind erosion, such as yardangs, may be evident on the stripped surface. River action removes the products of weathering and when these are deposited in topographic lows they will tend to be sorted into aeolian features. The lack of vegetation and soil allows individual rock bands to be well exposed and relatively more sharply defined than they would be in humid areas, and talus tends to be rare.

A good example of this type of landscape is provided by the Colorado Plateau of the south-west United States.

PART TWO · MAN AND THE DESERT

). Desert hydrology

ntroduction

Jp to now the emphasis has been on the physical characteristics of he warm deserts of the world; and for the purpose of delimiting hese areas some broad definitions of aridity in terms of climatic haracteristics have sufficed. But from the point of view of man's occupancy of these vast expanses of territory many other features nust be considered, for the factors which determine whether an area s inhabited or economically exploited are by no means entirely physical.

So, for example, certain types of settlement have been implanted n deserts solely because of the political and economic requirements of more densely populated regions. Obvious examples of this are centres of communication networks, airports, wireless and pumping tations, etc. Equally, mineral deposits, of which oil is probably the nost obvious example, may lead to some human activity in deserts n spite of the inherent inhospitality of the environment (Fig. 57). n this connection it should be noted that some mineral deposits, notably nitrates and phosphates, are particularly rich in deserts as the esult of concentration of salts under arid conditions.

The sea may also attract settlement to the coasts of some of the world's most barren deserts (see Fig. 58). Indeed, it is no coincid- nce that many rich fishing grounds, such as those off South West Africa (Namibia) and Peru, are associated with the same cold urrents which help to create the arid conditions of the littoral strip. Jpwellings of cold waters along the coast bring mineral nutrients nto the *euphotic zone,* the top layer of the sea where light penetra- ion is sufficient for photosynthesis to take place. The plentiful upply of nutrients allows good growth of phytoplankton, which orm the first link in the food chain that leads via zooplankton and mall fish to the commercially useful species that man fishes for. The ish not only provide a local source of foodstuffs and export income, ut in a dried form may even be used for fertilisers and animal food. Pearls, sponges, guano, seaweed and salt are also maritime resources hat have economic importance in certain coastal deserts. Littoral ettlement may be further stimulated in those places where the desert lies close to a major maritime route. No better examples can e provided than the various city states which developed at different imes to exploit the strategic and commercial advantages of the Red ea and the Gulf.

Some settlement may even arise from the inherent qualities of the desert environment itself. Deserts have always offered certain attractions as places of refuge where communities may live according o their own principles of organisation, as is the case with the Ibadhi community in the oases of the Mzab (Algerian Sahara). They have also acted as places for contemplation and seclusion (e.g. St.

57 The traditional economies of many Middle East states have been completely altered by the exploitation of oil. In Bahrain, for example, agriculture has declined in importance and many people find employment in oil refineries and in factories which utilise cheap oil and natural gas. In this photograph pipelines are crossing the sabkha (see p. 18) on their way to the Bahrain refinery.

Catherine's monastery in Sinai) and as centres of repose and recreation (e.g. the eighth-century Umayyad palaces built on the edge of the Syrian Desert). Today, the same advantages are stimulating the development of health and tourist resorts in desert regions, particularly now that the use of air conditioning has removed one of the major drawbacks of the environment.

One feature common to most of the settlement types described above is that their existence depends on contact with, or security from, the outside world. Thus the political and economic factors which govern accessibility assume overriding importance and the problem of water supply tends only to play a role in site selection rather than in the general decision as to whether a resource is to be exploited or not. Indeed in cases where the economic value of an activity is high, the potential returns may well justify shipping or piping water in from considerable distances or manufacturing it locally by desalination processes (see p. 77).

On the other hand water supply is of paramount concern for most indigenous societies and it is for this reason that in this short study the hydrological theme has been selected as the basis for understanding how and where man lives in the desert. But although the emphasis will be placed on the way in which water quantity and quality affect the traditional patterns of human occupancy based on livestock-raising and cultivation, this should not be taken to mean that these are the only factors governing desert land use even in such relatively simple societies. Quality of grazing, for example, obviously depends not just on hydrological conditions but on the way that plant communities in a particular area have adjusted to all ecological

58 In some coastal deserts, fishing and maritime commerce may form the basis of life. Dhows may travel from the Gulf as far as East Africa and India. This photograph was taken in the Gulf in about 1959.

factors. In agriculture, desert soil (see p. 22) and accessibility may also be of significance, but generally speaking such aspects only assume importance in explaining land use within hydrologically favoured areas and are of secondary importance when examining the broader aspect of where economic life is possible within a major desert region. In other words, selection of factors for study will depend on the size of the area to be considered. At the scale to be discussed here, water assumes overriding importance.

Classification of water supply

Concentrations of water in the desert may be classified on the basis of whether they originate from within the zone itself, in which case they are sometimes termed *endogenous,* or whether they originate from outside it, in which case they may be termed *exogenous.* These latter may in turn be subdivided on the basis of whether the flow is primarily in the form of surface supply (rivers, etc.) or a subsurface supply (*groundwater* flow). From the point of view of the actual volume of water available in deserts, exogenous supplies are much more important than endogenous supplies and by far the largest population clusters occur along permanent rivers which derive most of their supply from outside the region.

A further useful way of examining water resources in deserts is to consider them from the point of view of the nature of the drainage basin, and here a general classification proposed by de Martonne is often employed. This divides the land surface into basins with three types of flow: that where the flow eventually reaches the sea (*exoreic*), that where the drainage is into internal basins (*endoreic*)

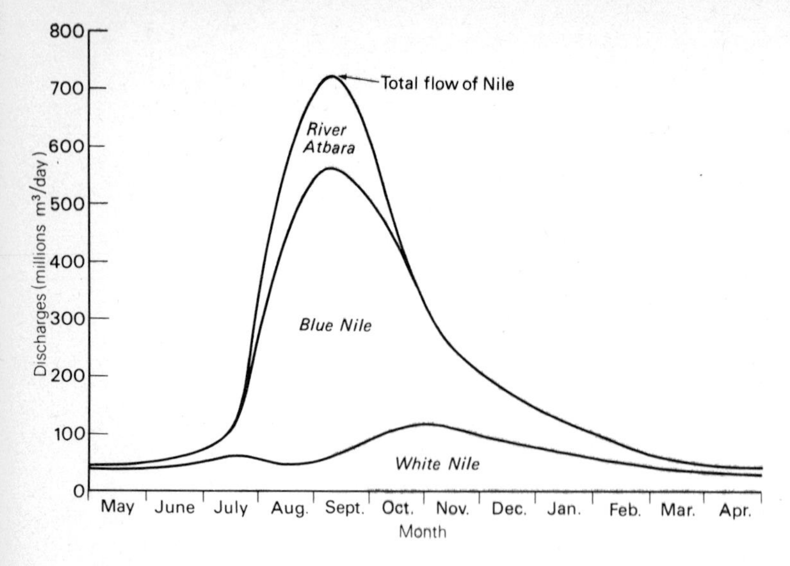

59 The discharge of the Nile where it flows through the arid tracts of Egypt is derived from wetter areas. The Blue Nile, with its highly flashy hydrograph, derives much of its discharge from seasonal rainfall in the Ethiopian mountains, while the White Nile, with a more evenly distributed discharge, derives its water from equatorial regions. This hydrograph is for the Nile at Aswan.

and that where there is basically no integrated network of surface flow (*areic*). Endoreic and areic basins generally characterise desert regions, but once again it is the exceptional exoreic or allogenic streams which rise in humid areas and can maintain themselves through the deserts to the sea that are of the greatest importance from the point of view of human occupancy.

Riverine settlement

Because the major permanent surface flows that occur in deserts originate in areas which have few common features other than the fact that they are essentially non-arid, and because some of these rivers flow into an inland lake while others eventually reach the sea, each has its own characteristics in terms of regime and morphology. This in turn means that the dependent communities have evolved their own set of solutions to such problems as drainage, salination, flood-control, lift, water-sharing, navigation and the like, so that hydrological organisation tends to be unique to each basin. As a result, few generalisations are possible in a book this length and so attention will be drawn only to a few points of particular interest.

The first of these is that co-operation between those who share the common water supply is absolutely essential. This, of course, is true in all *hydrological communities,* but in the case of a river where a misuse or a misappropriation of water upstream can have the most serious large-scale consequences for those living downstream, careful regulation of water-use is a *sine qua non* and some sort of political understanding must be achieved between all the riverine inhabitants over hydrological matters, even if such co-operation does not necessarily extend into other domains. It is no coincidence that many of the earliest advances in civilised living started in the river valleys of the arid lands.

The next point is that rivers draining into or through deserts normally have a markedly seasonal regime (Fig. 59). Consequently, the water in the river is considerably below bank level for part of the year and so some form of artificial lift (pumps or barrages) has to be employed if it is to be used for perennial irrigation. In early traditional irrigation practice, therefore, it was floodwater that tended to be the chief source of wealth, and thus the regularity of its rise assumed critical importance. This can best be illustrated by quoting the introductory remarks to an eye-witness description of a

disastrous flood failure in Egypt at the beginning of the thirteenth century:
'Eighteen cubits is the limit of the rise for the needs of Egypt. If it becomes more than this, high places become irrigated: there is then a superfluity. The highest limit of the river, occurring only rarely, is some fingers beyond twenty cubits; when this happens some places become submerged for a long time, and the proper period for sowing passes without their profiting by it . . . the minimum of the flood necessary is sixteen cubits . . . when the waters rise more than sixteen cubits to eighteen, all the lands accustomed to be inundated participate in the flooding and the crop suffices to furnish the needs of the country for two years or more. If, on the contrary, the flood remains under sixteen cubits, the part of the land inundated is insufficient: the crop does not furnish the needs of a year and there is a scarcity of food proportional to how far below sixteen cubits the waters stop. In the year 596 (end of 1199 AD) the river did not rise more than twelve cubits and twenty-one fingers . . .'

The less squeamish reader might profitably read the rest of 'Abd al-Latif al-Baghdadi's account for understanding the demographic consequences of a Nile failure in the period when basinal irrigation was the basis of agricultural organisation in Egypt. This may be found in K.H. Zand, J.A. Videan & I.E. Videan (transl.) *The Eastern Key* (Allen & Unwin, 1965).

The third point is that such river settlements form isolated centres of population. The Indus and the Nile, for example, both of which gain their discharge from humid mountainous belts, are lined by a relatively narrow thread of irrigated land and human activity. Such isolated centres thus exhibit many of the same characteristics associated with the smaller oases which use groundwater resources in a desert environment. The basic difference is really only that of scale.

Deep aquifers

Of less importance to settlement than the perennial rivers which rise externally, are the deep fresh-water aquifers underlying some desert basins. Such aquifers may have been charged during pluvial phases when the arid zones were less extensive than they are now (see p. 11) or they may still be recharged from relatively humid areas. However, since the nature of this kind of water supply is inherently interesting and since some knowledge of the mechanics of the supply is essential in order to assess its potential, it is necessary to give rather more attention to these aquifers than would be the case if water resources were being assessed purely on the grounds of the number of people they support.

The facts are fairly simple. Geologically many deserts have been formed by the deposition of sedimentary material in the down-warped margin of a continental shield, or by the infilling of internal structural depressions. As a result the rock strata often have a synclinal (basinal) or monoclinal (half-basin) form with the oldest beds outcropping on the margins of the basin. Where these beds are formed by alternating permeable and impermeable strata, water will enter the porous formations through the surface exposure and eventually work its way downslope to fill the basin. The rate at which it will move is referred to as the *transmission rate* and this will

60 Diagrammatic cross-section of the edge of an artesian basin.

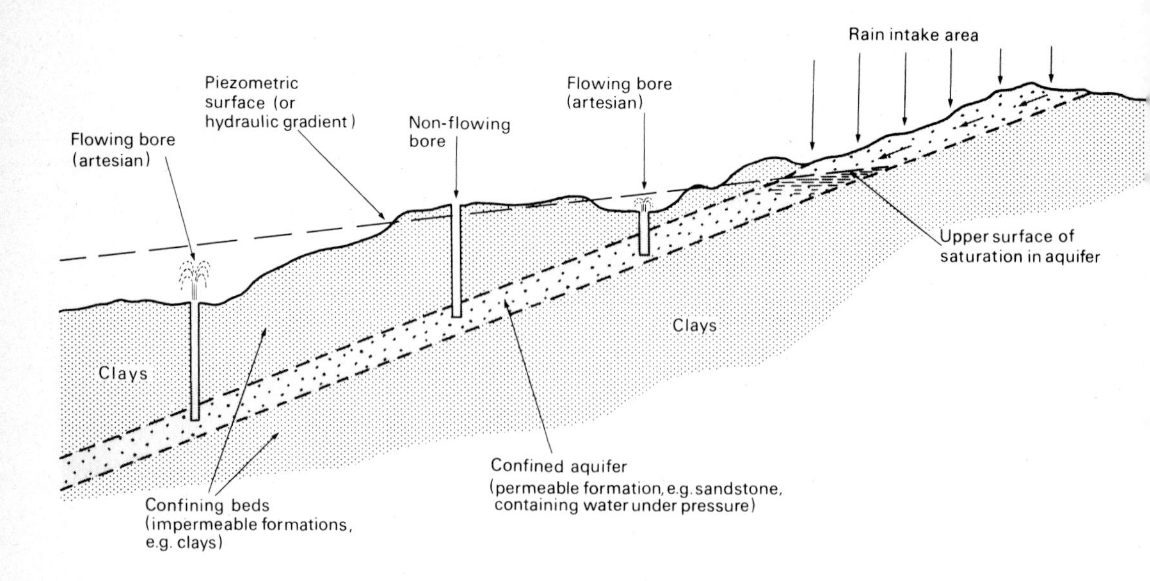

normally vary from about 2 m per day to 2 m per year according to the pressure of the head of water and the resistance of the formation. Since this water can escape neither upwards nor downwards it will remain confined and under pressure and its age will increase the further away from the area of recharge it is. (Techniques for dating water may therefore help determine where it originated and when.) Such a water-bearing formation is termed a *confined aquifer* and the hypothetical surface to which the water will rise at points of pressure release (i.e. where *hydrostatic pressure* is approximately equal to *atmospheric pressure*) is known as the *piezometric surface* (or *hydraulic gradient*). If, therefore, a well is sunk into the formation the water will rise in the bore and provide *artesian* flow (Fig. 60).

Artesian springs may also flow at points of natural pressure release. Thus, for example, most of the oasis settlements of the western desert in Egypt have been developed at those places where surface depressions (partly formed by wind erosion) have cut down towards the deep water-bearing formations in the Nubian sandstones that overlie the ancient basement complex or where the confined aquifer itself rises towards the land surface in a local anticline. In places this water may be 25,000–50,000 years old. Equally intriguing are the freshwater springs (long used by pearl divers) which discharge at points of local pressure release in the sea around Bahrain from the confined aquifers that originate in the scarplands of the eastern part of the central Arabian shield. Indeed the whole marvel of these artesian springs, which also water the oases of eastern Arabia and Bahrain Island, gave rise to an ancient mythology concerning the search for eternal life that was incorporated into the Sumerian epic of Gilgamesh, while the very name *Bahrain* (originally applied to the whole region but now only to the island) means 'the two waters', that is the water which is under the firmament and that which is above (Genesis I, 7).

The potential for extending the areas irrigated by such confined aquifers, however, is limited by four main features. The first is that the thickness of the individual water-bearing members interconnected within the general aquifer formation is very variable and it is a serious error to conceive of the whole as forming a continuous nappe of underground water. The second is that the quality of the water also varies, the salinity tending to increase downslope of the point of recharge; thus it is only under certain favourable conditions (either where the aquifer is particularly well-sealed from contact with sources of salt, or where rates of transmission are relatively high) that freshwater tongues extend far out underground into the desert. The third is that the water is often at considerable depth and so can only be exploited by expensive wells furnished with some form of mechanised lift; thus the costs of drilling, installing, maintaining and supplying wells and pumps in remote areas are often prohibitive unless subsidised by the state, as in the case of the new oil-rich countries, or by international aid programmes. The fourth involves the problems of recharge and requires a little more explanation.

The most important piece of information required for planning the exploitation of any aquifer is its *practical sustained yield,* that is the maximum amount of water which can be abstracted indefinitely without causing a critical drop of water levels. This is a function of its recharge rate and assumes a regular supply of incoming water. If recharge is not taking place then the water is entirely *fossil* and this means that any scheme for development must be formulated on the basis of maximising the return from an expendable resource.

Calculating recharge rates is difficult, for it requires knowledge both of where recharge is taking place and in what volumes. The former is less easy to determine than might appear, for in reality few confined aquifers conform to the simplified model outlined above and most receive some recharge through, and also discharge into, overlying formations as well as from the area of surface exposure. The volume is even more complicated to measure because it varies according to the climatic conditions at each point of recharge and these fluctuate in both the short and long term. Exploiting a newly discovered underground aquifer is therefore a complex process and requires a considerable amount of preliminary research aimed at discovering the location and volumes of present-day recharge and discharge and the transmission characteristics and salinity parameters at points of potential exploitation. Optimising the yield in turn may involve vast expenditure in maintaining pressures and preventing the incursion of saline waters from other aquifers or the sea.

Fig. 61 shows the scale of the area over which study must be carried out in order to plan for a project such as the New Valley scheme in Egypt (as distinct from the Old Valley, i.e. the Nile). In this particular case the evidence points to the fact that some recharge may still be taking place from the direction of the tropical plateaux in north-east Chad (where the rainfall ranges from 250–625 mm/year). But it has been calculated that the total recharge from the borders of the Sahara into all the confined aquifers which underly this desert of nearly three million square miles is only 4 milliard (4×10^9) m^3 per year; that is somewhat less than 5% of the average annual flow of the Nile. In the case of the Eastern Arabian aquifers the recharge areas are now within districts of very sparse rainfall and the recharge is negligible.

Thus, while a study of the characteristics of confined aquifers in

61 In Egypt there are plans to utilise groundwater for the New Valley scheme to the west of the Nile. Groundwater in the area is in part derived from high rainfall areas in Chad.

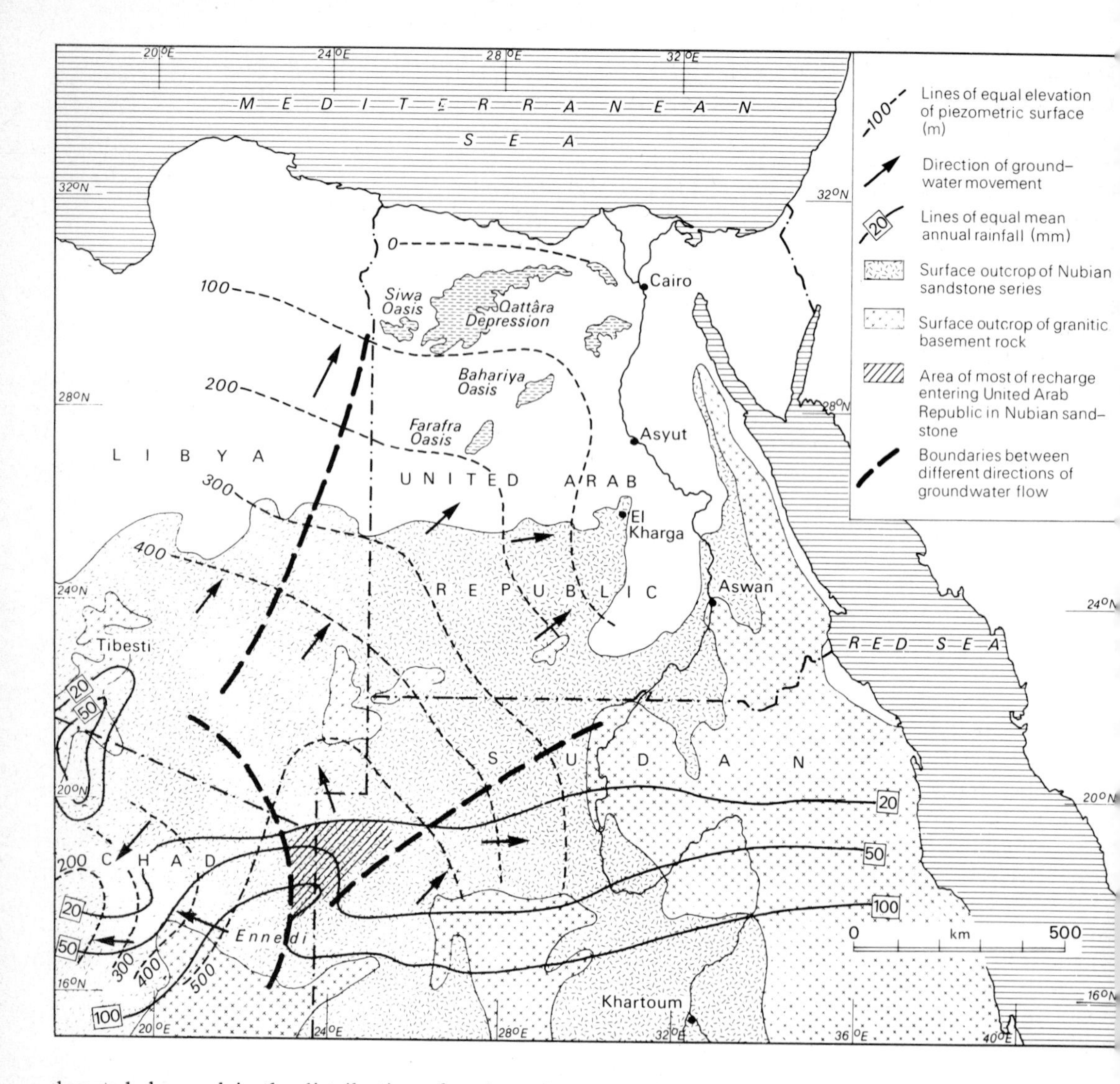

deserts helps explain the distribution of oasis settlement, and while the problems posed in exploiting them are inherently interesting, the fact remains that their potential is much more limited than many sensational reports would have one believe. Even if the technically and economically impossible were to be achieved and every drop of water stored in the seven great basins of the Sahara were to be extracted, the water would be sufficient for 150 million people for 500 years. That is, it would only allow the present population of approximately 150 million in the thirteen states which have a territorial interest in this desert, to double, an increase that in any case will occur within the next 25 years or so. Looked at in another way all the water storage is simply equal to the amount of water that the Amazon discharges into the sea every 2 years. Therefore, when reading about schemes for utilising aquifers containing fossil water it is necessary to keep them in perspective.

Endogenous water supply under present climatic conditions

It has already been shown that precipitation within deserts is highly variable in space and time. In general, the lower the average total the greater the variability (Fig. 3). Thus precipitation within deserts is an unsatisfactory source of supply in terms of quantity and reliability, and both drought and flood are recurrent problems. As Isiah Bowman wrote in his *Desert Trails of Atacama*: 'It is the fate of desert communities that they should be devastated by the same agent to whose gentler operations they look with such delight.'

When severe storms occur in the desert there is very little vegetation to intercept the precipitation, and so very little *interception storage* occurs. The bulk of the rain falls on the soil surface. Some will infiltrate, though as shown on page 40 the properties of many desert surfaces are not conducive to a very high rate of *infiltration.* If the rainfall rate is greater than the *infiltration capacity* of the soil the water will begin to flow as a sheet to fill the minor depressions on the ground surface (*depression storage*). As the flow begins to increase it will tend to accumulate in small rills which lead through ever larger channels until they form a temporary river or *wadi* (see p. 40). Thus soil type, surface conditions and relief are major controls of the location and quantity of endogenous water supply in the desert.

Water accumulation in a desert platform: desert lakes

On the more or less flat surface of a desert platform the only soils that tend to develop extensively are silty with fine particles. So fine are the particles that in some cases the surface areas of grains contained within 1 cubic metre of soil may be as much as 35 hectares. When it rains, the initial infiltration rate will tend to be high, at 10–20 mm/h, but because the ratio of pore space to solids is low and the clay particles swell by *adsorption,* the soil surface rapidly becomes sealed into a pan and water flows over it to accumulate in hollows, where it is eventually evaporated. Unfortunately, as it evaporates salts are precipitated, so that even though a small lake may form from time to time in such a depression, the water in it soon becomes undrinkable because of the build-up of salinity. As a result, most of the rainfall that accumulates in flat desert areas is rendered useless for consumption by animals or man.

However, two important exceptions occur: one is where the surface of a platform is made up of limestone, the other is where it is covered by sand dunes.

Water accumulation in a desert platform: limestones

In areas where the rock surface is formed by a limestone some of the water that falls will infiltrate into the rock, because limestone tends to offer fissures which allow its passage. Under present climatic conditions the volume of water accumulating in this way is normally too small to develop any major *karstic* features, but sometimes little solution hollows, with freshwater pools in them, do occur. In areas of higher rainfall, as at the great classical site of Palmyra in Syria, karst springs may form. In full desert conditions, however, water

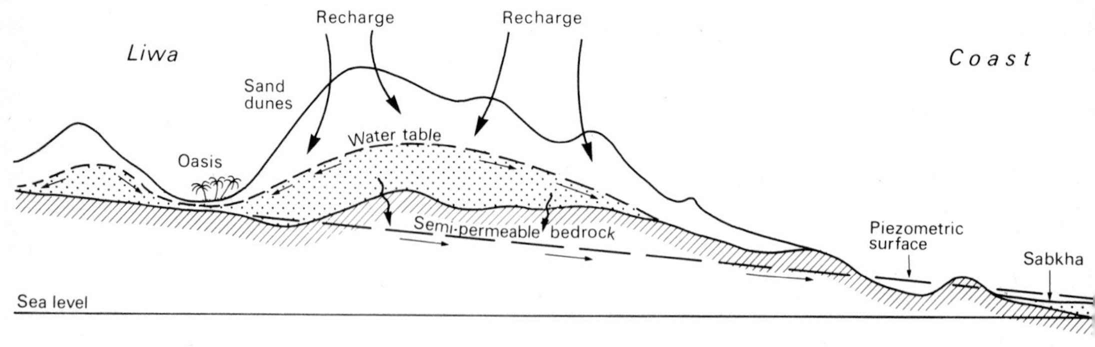

62 A schematic cross-section (not to scale) of hydrological conditions between Liwa and the coast in Abu Dhabi. A perched lens-shaped freshwater reservoir occurs in the dunes above semi-permeable bedrock.

accumulation in limestone is normally inadequate for cultivation, but the type of depression generally referred to as a *daya* which characterises some limestone desert surfaces, may provide a semi-permanent supply adequate for watering flocks, and even a little seasonal cultivation of cereals (see Fig. 30).

Water accumulation in desert platforms: sand

Sand dunes, the nature and form of which we have already referred to (see p. 32), provide the one major reservoir of water which permits man to live far out into the desert. This is because the sand itself is both porous and permeable, and because the dune height permits the water to accumulate below the zone of intense evaporation. It is worth remembering that capillary rise, even in fine sand, is unlikely to be more than about 1 m per year, and that intense heating will not occur to any great depth, sand being a poor conductor. Consequently evaporation losses from the capillary fringe will be far lower than with other desert soils.

When it rains the water infiltrates rapidly, and, provided the underlying platform is more or less impermeable, accumulates in a dome-shape. Considerable volumes of water may be held in this way in large dune areas, and it is even possible for lateral seepages from the perched water table above hollows to be sufficient to water a few palms. This is the situation in the case of the Liwa Hollows of Abu Dhabi shown in Fig. 62. Normally, however, the water reservoirs are smaller and only adequate for small nomadic groups such as the bedouin. Even then, it must be emphasised, these freshwater reservoirs only occur when rainfall is fairly regular: in the main part of the Empty Quarter of Arabia, for example, dunes are normally completely dry because of the conditions of extreme aridity.

Settlement on the coasts of flat desert areas is also sometimes possible due to the small accumulations of water which form in the dunes that develop along the shore. This freshwater reservoir will tend quite often to have a lens-shape as shown in Fig. 62. Since the principles which account for the formation of such lenses are the same as those which affect freshwater occurrences in any coastal area, this will be an appropriate place to discuss the whole question of what happens to freshwater when it comes into contact with a saline water body.

Freshwater on the coast

Freshwater has a lower density than saltwater, such that a column of seawater can support a column of freshwater approximately 2½% higher than itself (or a ratio of 40 : 41). So where a body of freshwater has accumulated in a reservoir rock which is also open to penetration from the sea, it does not simply lie flat on top of the saltwater but forms a lens, whose thickness is approximately forty-one times the elevation of the piezometric surface above sea level. This is called the Ghyben–Herzberg principle (Fig. 63). The corollary of this rule is that if the hydrostatic pressure of the freshwater drops due, say, to over-pumping in a well, then the underlying saltwater will rise by 40 units for every unit by which the freshwater table is lowered.

It is this phenomenon which explains the extreme delicacy needed in the exploitation of freshwater near the coast, and why it is that near the sea a change in the hydrological conditions high up an aquifer is most strongly felt. Two examples will help illustrate this point. The first will be taken from the case of the confined aquifer of eastern Arabia which irrigates the palm groves of Bahrain, an offshore island. At the north-western end of the island the hydrostatic head is still some 3.35 m above sea level, but due to over-exploitation this is now dropping at a rate of some 10 cm per year. The saltwater entering at the seaward end of the aquifer is therefore spreading up the aquifer with a vertical rise of 4 m or so each year

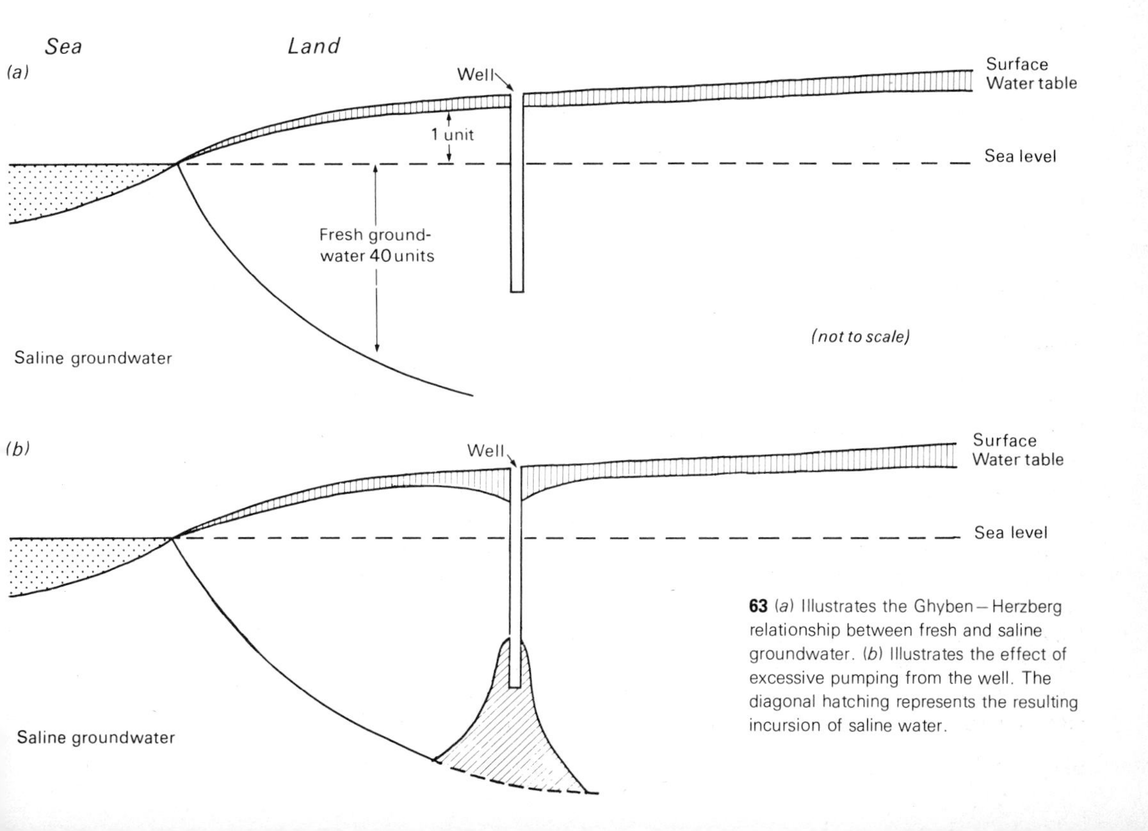

63 (*a*) Illustrates the Ghyben–Herzberg relationship between fresh and saline groundwater. (*b*) Illustrates the effect of excessive pumping from the well. The diagonal hatching represents the resulting incursion of saline water.

and this means that even if the inhabitants of the northern coast are not themselves yet paying for their prodigality, those on the eastern coast are.

A similar situation arises in the case of the Nile delta where the groundwater is *unconfined,* that is the upper surface of the water table is free to rise and fall with changes in the hydrostatic pressure. As a result of constructing the Aswan High Dam, groundwater levels have dropped downstream and so near the coast saltwater has intruded and, because the aquifer is unconfined, has salinated the soil above, thus rendering it useless for cultivation. This is one of the costs which have to be paid for the undoubted gains brought about by rationalising the flow of the Nile so that little of the water is wasted by draining out to sea.

Even more delicately balanced is the water potential of the small dune lenses in coastal deserts with which this discussion started. This is partly because the lens forms an isolated body surrounded by seawater, and so is subject to the turbulence caused by tidal action on all its sides (this renders the lower fifth of the freshwater body very brackish) and partly because the recharge only comes from sporadic rainfall and even this tends to enter the reservoir in a somewhat brackish state due to downwashing of the marine salt particles deposited on the surface of the sand. But even though the total volume of freshwater held in coastal dunes is never great and the quality falls off rapidly after every storm, these lenses may be sufficient to support a small coastal community. Such water supplies originally gave rise to the nucleus of five of the seven Trucial States (now called the United Arab Emirates). Furthermore, because the freshwater forms lens-shaped bodies which extend below sea level, it is sufficiently thick for trees to take root, and it is this which explains the perhaps rather surprising fact that in an absolutely barren desert a fringe of palms may sometimes be found along the edge of the shore: indeed one curious aspect of the phenomenon is that under the mirage effect these trees may even appear to be growing directly out of the sea or salt flats.

Wadi basins

Up to now this discussion on the formation of natural freshwater reservoirs in a desert has been confined to essentially flat areas where run-off is inhibited by lack of relief and local drainage networks are poorly developed. To understand how an integrated drainage basin can develop in a desert region and to appreciate what happens to the surface and underground flow within it, it is necessary to introduce the element of relief. The basic model is one in which the wadi rises in an area of considerable uplift and runs out into a flat area such as we have been describing. For convenience these two areas will be termed the mountain zone and the desert platform.

The run-off in the mountain zone

The run-off in the mountain zone will be rapid for the slopes are steep and devoid of any vegetation and soil. Flows therefore occur easily, but in storms of low intensity a high proportion of the water will enter the coarse detritus of the wadi bed. In larger catchment areas the water stored in such shallow aquifers may maintain the

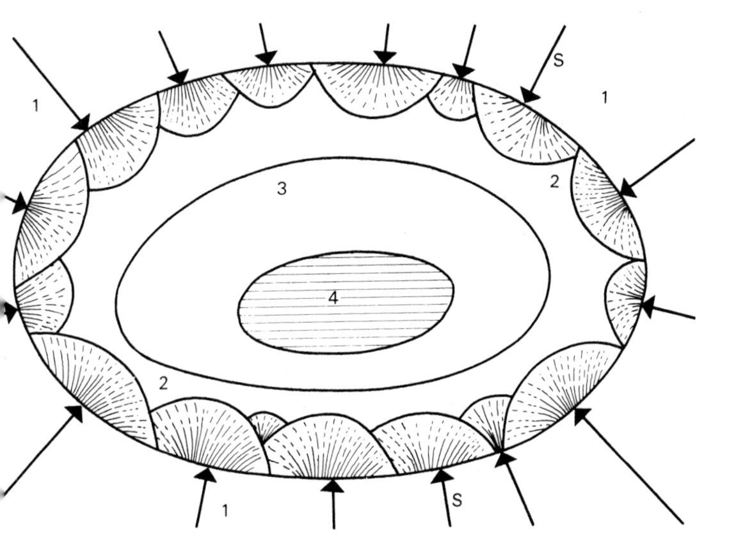

64 Schematic plan of an endoreic basin. 1, Mountainous zone with gully and sheet erosion, producing some substantial and occasionally permanent streams (S). 2, Bajada zone of contemporary and fossil alluvial fans. 3, Desert platform of low gradient and generally fine sediment. 4, Playa zone, periodically flooded.

flow on or near the surface of the gravels for quite a time wherever a rock bar or similar feature constricts the wadi channel. In cases of prolonged or high intensity rainfall, however, the channel will fill so that surface flow will be carried out of the mountain zone into the *bajada* zone (Fig. 64). Such flow, as we have already seen (see p. 40) fluctuates rapidly.

The bajada zone may be defined as the zone of coalescing debris fans, formed as the streams deposit their load where they spread out over the desert platform from the confines of the mountain course. It is thus a transitional area whose considerable hydrological importance will be better appreciated when we consider what happens to the groundwater rather than the surface flow in the drainage basin.

Groundwater flow in a wadi basin

Within the mountain zone much of the water flow is ephemeral. *Base flow,* that is sustained perennial flow, derives from two sources: the discharge of aquifers that may have developed within the parent rocks, and the shallow aquifers which develop within the gravels and sands of the wadi deposits themselves. The importance of the former depends on local lithology and morphology, and so, in order to simplify the discussion it will be assumed that the parent rock is more or less impermeable and that base flow derives solely from water stored in the fluvial deposits.

Because the wadi valley within the mountain zone is narrow, the detritus coarse and the profile steep, the volume of stored water is low (except in local basins) and is fairly rapidly drained downstream. Thus base flow rapidly falls off in the dry season, and this, combined with the fact that land suitable for cultivation is difficult to find in such terrain, explains why the possibilities of perennial irrigation in this zone are normally very limited.

By contrast the volume of water held in the *piedmont* zone is high because here the surface and subsurface flow of the upper catchment which has been concentrated at the head of the fan suddenly spreads out and sinks into the mass of fluvial detritus that has been deposited at the point where stream velocity is reduced. On the other hand this does not mean that underground flow in the bajada zone ceases altogether; rather that its nature changes. Now the groundwater, hitherto laterally confined, coalesces with the discharge of other wadis and so moves out under the desert platform in a slowly moving broad front of increasingly saline water. Because the lithology of the desert platform is frequently different from that of the mountain zone this water may sink to a considerable depth before it is eventually stopped by an impermeable formation (*aquiclude*). If in the process it passes out under another impermeable formation then the water will become confined and give rise to the type of relatively freshwater aquifers that have already been discussed. Normally, however, it will find its way out to sea, or into the centre of an internal drainage basin where it once again comes near the ground surface and evaporates in a playa, thus completing the local hydrological cycle.

Freshwater occurrences downstream of the bajada zone in a wadi basin

Although a wadi tends to lose its identity downstream of the bajada zone, some flow may continue out into the desert platform following the old drainage lines: here too it may receive supplementary flow from local run-off. Although there may be no surface or subsurface flow in these channels for long periods, the various fluvial deposits (gravel trains, fans, calcretes, etc.) and sub-basins each have their own hydrological characteristics and in some the groundwater flow may be adequate to provide wells for a small nomadic group for part or even all of the year. Furthermore the differences in soil moisture characteristics are also reflected in a range of seasonal and perennial vegetation communities, each of which has its own grazing potential (see p. 26).

Dew and fog

Finally, by way of completing this chapter on the hydrological resources of the desert, a few words should be added on the subject of dew and fog. Generally speaking, the former adds little if anything to water storage because even near the coast where the total volume precipitated in a year may be relatively high (say up to 40 mm per annum) the maximum dew formation recorded for any one night is less than 0.4 mm and this is rapidly evaporated during the day. Vegetation may benefit indirectly through a temporary reduction in transpiration, but its value to direct growth is normally very limited indeed. Animals grazing on dunes covered by vegetation can also receive a certain amount of free water in this way. By way of curiosity it might also be mentioned that there is a tribal group which lives on the south coast of Oman in an area without wells, reputedly managing by hanging up blankets to collect the dew that forms at night under the particularly humid conditions which prevail there in summer.

The same strictures apply in general to fog, but in this case it is clear that under those exceptional circumstances in which fog forms regularly, the moisture can play a role in determining vegetation forms (e.g. the *loma* vegetation which grows in summer on the coastal ranges of the Peruvian desert).

7

The hydrological factor in desert economies

Water quality

The previous chapter described the principal types of water resource in desert areas. Before we can understand how man has exploited them, it is essential to look at the problem of water quality, for living in the desert is as much a matter of salinity tolerances as it is of economising in the volume of water consumed. Indeed, because few desert areas are without access to water of some kind or other, salinity may be considered the principal factor that limits the potential for settlement; if the dissolved salts could be removed easily and cheaply then the major problem of the arid zones would be solved. Why is it that there is a greater salinity in the arid zones than there is in the humid?

Salt accumulations

The answer to this question lies in the fact that salts are concentrated rather than dispersed. There are three principal reasons for this:

(1) *Evaporation.* In the dry, hot and windy environment of the desert, the tendency is for any moisture in the regolith to be drawn upwards so that the salts are concentrated in the upper soil layers rather than leached out of it as in more humid regions, a point which has already been made with regard to calcrete formation (see Fig. 11). This process poses particular problems for irrigation in areas where the water table is high. For example, in Iraq it is estimated that some 3 million tonnes of salt are added to the cultivated land each year even though the irrigation water from the Tigris and Euphrates is itself sweet (approximately 200–400 parts per million of total dissolved solids; see below).

(2) *Endoreic drainage.* The arid zones tend to be areas of endoreic drainage and so salts are not evacuated out to sea as is the case in areas of exoreic or through-drainage. Following on from this salts are readily drawn up to the surface because water tables may also be relatively high in areas of inland drainage.

(3) *Atmospheric recycling.* This introduces salt into areas that might otherwise be free from the salinity hazard and may even bring in salt from external sources. It largely results from winds which pick up salts from surfaces of evaporative bodies such as the sea, salt lakes, sabkhas and playas, but in maritime environments rain and fog may also precipitate salt particles.

Finally, while the rocks in deserts are not necessarily inherently saline, the proportion of evaporites increases the longer arid conditions have prevailed.

Table 10. Water salinity characteristics

Nature of water use	Concentration of salts (ppm TDS)
Sea water	35,000
Maximum potable level for man	3000
Recommended potable level for man	less than 500–750
Domestic animals of desert	less than 15,000; extreme 25,000
Irrigation water (assuming optimum soil and drainage conditions)	less than 750, no salinity risk
	750–1500, restricted yields of very sensitive crops
	1500–3500, restricted yield of many crops
	3500–6500, salt-tolerant crops only
	6500–8000, restricted yield of salt-tolerant crops

Measures of salinity

Obviously not all salts are as harmful to life as others, and, indeed, small quantities of some are even essential to it. Furthermore, different minerals concentrate in different places and this of itself may lead to significant variations in land and water use. But, broadly speaking, it is the total volume of dissolved salts in the water which defines its *quality* and provides the general parameter which indicates such particular features as the sodium hazard to cultivation.

This salinity may be measured directly, in which case it is usually quoted as TDS (*total dissolved solids*) in ppm (parts per million) by weight of dissolved matter in a million parts of water, but sometimes as mg/l (milligrams per litre). Alternatively it may be measured indirectly through the *electroconductivity* (EC) of the solution. This is easily determined on a conductivity meter and expressed in milli- (10^{-3}) or micro- (10^{-6}) mhos (the reciprocal of that measure of resistance called the ohm), usually at 25°C. (The conductivity of a solution depends on its temperature and on the nature and quantity of dissolved materials in solution.) Although EC does not give a precise determination of the total dissolved solids it does give a clear and easily measured indication of the degree of salinity and is thus much used.

Since the salinity of desert groundwater may range from less than 300 ppm of TDS to over 1000 times that figure, a useful point of reference may be provided by remarking that the average figure for seawater is around 35,000 ppm, 85% of which is made up of sodium and chloride ions (Na^+Cl^-).

Hydrological limitations to life in the desert

The limiting factor controlling where and how man can live in the desert is man himself, because he is basically not physiologically adapted to the environment. He needs to drink frequently and his total consumption in a year is some 50% more than a camel. The maximum salinity he can tolerate for any length of time is of the

Table 11. Water consumption in different types of settlement in eastern Arabia

Type of settlement	Consumption per head (litres/day)
Very small desert settlement (e.g. fishing) with traditional water supply	28
Very small desert settlement with government water truck supply	80
Domestic requirement – traditional agricultural village	120
Domestic requirement – agricultural settlement planning figure	160
Town (no major industry) traditional supply	160
Town (no major industry) planning figure	240
Modern oil-boom town (no industry)	400–1800

order of 3000 ppm TDS, but this figure should be compared with the 500 to 750 ppm recommended by the World Health Organisation and the United States Public Service (Table 10).

Furthermore, man tends to 'waste' water and he obeys a kind of Parkinson's law, in that his capacity to use water expands to meet the water available. This may be seen in Table 11, which gives water consumption in different types of settlements in eastern Arabia.

Obviously the possibilities for an agricultural economy in the desert are limited. Whereas the natural vegetation makes extraordinary use of soil moisture and can adapt to drought and high salinities (see p. 24 for a discussion of xerophytes and halophytes), all cultivated plants require large quantities of relatively high quality water. The extreme limits for agricultural settlement are more or less determined by the requirements of the date palm, a remarkable tree which might be regarded as the agricultural equivalent of the camel in the desert regions of the Old World. It thrives in dry heat and is highly salt-tolerant with an extreme limit for economic viability of the order of 8000 ppm TDS. But, while the date palm compared with other crops is highly tolerant of salt, it is very demanding in the quantities of water it needs, and this is its limiting feature for use in the desert environment. Just how limiting can be appreciated by comparing the minimum water consumption of a bedouin group living at subsistence level on livestock herding with a similar group living at the same level by cultivating the palm: the latter will consume almost a thousand times more water than the former and furthermore the palm will require a salt content that will have to be some 50% lower!

Livestock herding

By far and away the most profitable use man can make of desert land is, therefore, to herd livestock. All the domesticated animals of the desert drink relatively small quantities of water (about 1 m^3 per annum) and tolerate high salinity. It has been observed of sheep in the Australian desert, for example, that they can tolerate up to 25,000 ppm NaCl and that some actually appear to thrive on salinities of 19,000 ppm. Yet long-term experiments also show that while 10,000 ppm had no ill effects on any of the herds studied,

65 The camel is the domesticated animal most highly adapted to the desert environment and can be used in terrain where neither sheep nor goats could survive. The main adaptations of the larger animals to the environment include: the ability of body temperature to rise above normal thus reducing the need to evaporate water to keep temperatures constant at 37°C; and kidney adaption which cuts down the loss of water in urine and also permits the animals to drink salty water.

15,000 ppm was detrimental to some and all suffered at 20,000 ppm. Furthermore those sheep which thrived at 19,000 ppm were on green feed, but when they were placed on dry feed their salt toleration fell to some 10,000–12,000 ppm.

So it can be seen that the factors which limit the possibilities for livestock herds cannot be determined simply in terms of quantity and quality of water alone, for the nature of the grazing is equally important and this in turn can affect the water requirements of each breed of animal. Even harder to define are the points at which the animals' economic value starts to decline, for although the domesticated animals of the hot deserts are capable of withstanding extremely hostile conditions, their economic yield is best when grazing and freshwater are plentiful, and declines as the tolerance levels for each breed are approached. Perhaps the easiest way of understanding how the desert land is used in nomadic societies, therefore, is to examine what the inhabitants actually do through a study of certain bedouin practices in the Middle East.

Nomadic grazing practices

In summer, the camels (Fig. 65) grazing on the outer fringes of the desert, are normally watered every 2 days, even though they may be capable of going without water for a week when feeding on the very good vegetation which occurs in some wadi beds. In full desert conditions, where the grazing is predominantly halophytic, they are watered daily. By contrast, the camel can go without any free drinking water at all in winter if the rains produce good fresh grazing (water content

70–80%), and under these conditions it may range as much as 25 km a day as it browses. Since it is then no longer tied to its water supply and appears to have no territorial instincts, this presents problems of herding: as a result camels in a state of water independence have to be hobbled (they are never penned).

Thus in winter it is man's rather than the camel's adaptation to the environment which limits his use of the grazing resources of the desert. In summer, on the other hand, the picture is quite different. Camels automatically return to water and spare man the hardships of herding at a time when physical conditions are at their most trying. At this season the grazing potential of the true desert is extremely limited: this is not just because there is little vegetation and water, but also because the area over which a camel being watered every day can range, is only a quarter of that which it covers when grazing permits watering every other day. Much of the area is thus deserted and in this season the camel-herding nomads either withdraw into the more favoured areas on the fringes of the desert or into the sand desert, which has superior water resources to other types of desert platform surface (see p. 62). It is this '*pulsatory*' nomadism which characterises the pastoral groups living in the desert. It is sometimes termed *horizontal* nomadism by contrast to the *vertical* displacement of nomadic groups which use the different seasonal pastures in a mountain zone, such as the Zagros Mountains of Iran.

The herding of sheep and goats gives rise to a very different pattern of land use. In the case of sheep, suitable pastures simply do not exist in full desert conditions because the sheep are essentially grazers and not browsers. Moreover, they need watering daily and their range of movement, 15 to 20 km, is more limited than that of a camel. To some extent this is also true of goats, but the facts that they are browsers as well as grazers, that they can survive water deprivation more easily and are generally hardier and more intelligent than sheep, does mean that they are better adapted to living in the desert and are ideally suited to arid mountain pasture. One thing both sheep and goats have in common, however, is that they only lactate when there is ample fresh grazing, whereas the camel normally lactates during 11 months of the year.

Land-use patterns of different socio-economic groups

Two important points concerning land utilisation and socio-economic organisation in the desert follow from these considerations. The first is that sheep and goat herding is virtually confined to the semi-arid borderlands in the vicinity of settled areas. Since sheep and goats, when grazing on natural vegetation alone, cannot provide a basic foodstuff by which man can live throughout the year, they are either kept by groups which associate with settled peoples, or by pastoralists who also graze camels.

The second conclusion is really the converse of the first; that camel-herding by itself can just provide a viable living, but that if the animal's adaptation to the desert environment is to be exploited fully, then it is not possible to keep sheep and goats as well. For this reason the occupants of the sand desert are pure camel-herders, and because water and grazing resources are scarce and widely dispersed, the territory of their tribal groups is extensive. This is also the basic reason why there are marked differences between the behaviour and attitudes of the camel-herding tribes and those which keep sheep and goats.

The frontier of settlement

Because desert animals are relatively salt-tolerant and economical in the volumes of water they require, and because they can feed on natural vegetation which utilises soil moisture with extraordinary efficiency, some sort of livestock-herding must be the basis of economies in the desert environment.

Thus three principal zones of land occupancy can be identified in desert regions: that where seasonal rainfall permits dry-land farming; that where grazing of domestic animals is possible; and that where no regular use can be made of the land. These correspond roughly to the climatic definitions of semi-arid, arid and extremely arid (see p. 2). However, there are no rigid boundaries between these zones. Large areas of quite useless desert may exist within the semi-arid zone, usually because of unfavourable relief or lithological factors, while extensive areas in the same zone may be better suited to grazing than to cultivation. Equally, nomadic groups may occasionally be able to penetrate right into the cores of deserts in search of grazing after exceptional rainfalls, when ephemeral plants (see p. 24) sprout. Nor is agriculture confined entirely to semi-arid zones, for irrigation may be possible well out into the desert by making use of some of the freshwater resources discussed in the previous chapter. Before describing some of the special techniques employed in pushing forward the frontier of settlement it is important first to emphasise that settled life in the desert is not only a matter of technical ingenuity and organisational ability, but also of compromise between the sedentary and nomadic groups.

Traditionally there is always a potential for conflict between nomads and settled peoples and this is most liable to build up at the frontier between the arid and semi-arid zones. Normally this potential is neutralised though a degree of social assimilation, demographic equilibrium, compromise over land-use, and economic exchange, but when the interests of the urban-based central governments which control the settled lands clash with those of the tribal organisations which dominate the desert, the equilibrium is upset and it is the frontier cultivators who inevitably suffer most.

Settled communities living in the desert itself, however, are in the full domain of the nomad and have to come to terms with their nomadic hosts if they are to survive. The equilibrium between the two societies is precarious for the pastoralists tend to take a short-term view of economic advantage whereas the cultivator of tree crops must have long-term security. Isolated oasis settlement, therefore, was and still is basically fragile. The cultivators not only have to keep the desert at bay but also its sometimes rapacious inhabitants.

These remarks, of course, apply to the traditional scene. Today the situation is changing. Modern communications and weapons have immeasurably strengthened the power of central government and tribal society in the Middle East, for instance, has more or less broken down.

66 An irrigation channel, cut into hard igneous rock, leading water from semi-permanent flows in the gravels of an Oman wadi. The gravels were formed by the constriction of an igneous rock bar upstream of the palm groves (middle distance).

Yet because of such past conflict, often rather inexactly described as that between the 'desert and the sown', nomadism is today frequently looked on with disfavour by modern governments and the tribesmen are encouraged to settle and cultivate. This is a pity because one of the results is that land which is suitable only for livestock raising is being abandoned, and that at a time when the pressure on rural resources is growing and the demand for animal protein developing. Surely, therefore, it is necessary to reappraise this policy and to examine how animal husbandry may be improved. Perhaps the answer lies in improving or developing breeds specifically for meat, wool or milk products and by integrating the two rural forms of life more closely, notably in the production of fodder crops. At the same time central government would have to provide the marketing services necessary for specialised herding. But before this can happen entrenched prejudices must be broken down.

Oasis cultivation

The techniques available for the exploitation of desert water resources for cultivation basically divide into two groups: those for exploiting surface flow, and those for exploiting groundwater. In the following sections some of the more interesting techniques traditionally used in the deserts of the Old World are described.

Catchment management or runoff-farming

Three sets of techniques were developed in antiquity which permitted harvesting of enough water from the scanty and irregular flows in a catchment area to ensure that a crop could be grown each year.

The first technique was that generally employed within a dissected highland region where the problem was often less that of collecting water than of developing the land itself so that regular natural flows might not be wasted. This might involve the laborious construction of terraces, clearing the land of stones and boulders, and creating a soil. Channel networks would also have to be created to direct the water to the point of use and this might involve diverting the natural flow by a long conduit, carved into the rock (Fig. 66) or built up along the side of a wadi course crossing obstacles by means of tunnels, inverted syphons and bridges.

The second set of techniques was that developed to utilise the flows which can occur at two main types of location within the upper part of a catchment: where a major valley has developed within a plateau area, and where a wadi breaks through the confines of its mountain course. In both situations the basic system involves diverting the flood so that both water and soil are deposited in the required area. This may be a simple matter of building a small dam or bund of stones, or some more complex structures may be required. The problem common to all such schemes, however, is the silt build-up from the heavily charged water flows (see p. 41). The silt, has, of course, considerable value in continuously supplying soil nutrients and in storing moisture, but it also means that the levels of the channels and dams have to be built up so that eventually the site of exploitation has to be shifted.

The management of small catchments that develop in areas of relatively low relief involves a third set of techniques, for here the flows are too small and too irregular to be of use without first improving run-off. In the Negev of Israel these ancient techniques have been studied in some detail and experiments in reconstructing ancient run-off farms show that even in a year when the rainfall is as low as 25 mm (and 250 mm is the generally quoted figure for the limit of dry-land farming in the Middle East) sufficient water could still be collected to produce a crop. The secret of this success lies in dividing the catchment area into small subunits and clearing the soil of stones in each so that the maximum amount of silty or loessic soil is exposed. Since this soil becomes impermeable very rapidly when wetted (see p. 61), a small shower will produce surface flow. This is then concentrated by means of artificial rills and channels towards a reservoir which feeds the cultivated area. In such micro-catchment areas the water collecting efficiency can be up to ten times greater than the natural concentrations which develop in a macro-catchment area. Modern applications of these techniques today may well have potential not only for cultivation but also for improving vegetation for grazing.

Storage

Associated with these techniques for improving run-off, were arrangements for collecting and storing water for both human and animal consumption in cisterns and reservoirs. The different forms these took are too numerous to discuss in detail here, but one of the most remarkable groups is the ancient tanks built over the

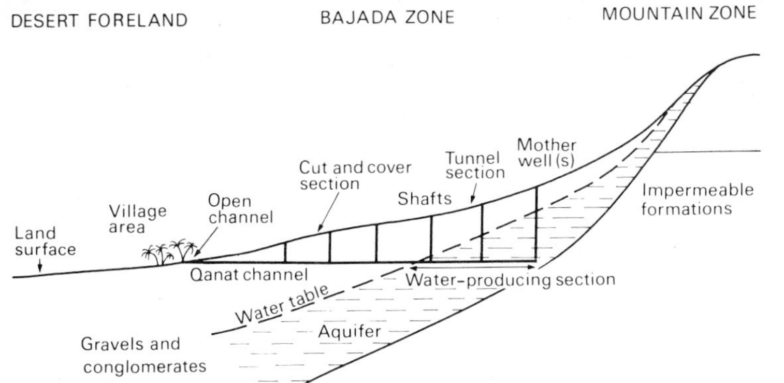

67 Cross-section of a Qanat in a piedmont zone.

68 An aerial photograph of a qanat landscape. The craters are the mouths of the access shafts to the tunnels where the spoil of excavation and clearing accumulates. The various crater lines indicate the different points where the aquifer has been tapped (the mother wells).

course of many centuries to collect the run-off in the Crater area of Aden. Shortly after the British started to reconstruct the abandoned system in the middle of the last century, it is reported that more water was collected in one day than all the wells could yield in a year.

Qanat

All the techniques described so far have been associated with exploiting surface flow in arid zones. Another set involves making use of groundwater. Perhaps the most remarkable of these is the system frequently referred to as qanat (pronounced kanāt), though many other names have been used, e.g. kariz, foggara and falaj. These were developed primarily to use piedmont water resources.

As can be seen from Fig. 67 a qanat is essentially a horizontal well in which the discharge of a reliable aquifer is brought to the surface by a tunnelled conduit. Just what constructing such a system involves may be appreciated when it is realised that even on a fairly moderate qanat the mother well may be 20 m deep, while the tunnel, to which a shaft has to be sunk from the surface every few metres in order to provide the builders with air and the means of extracting the soil may be several kilometres long. It is these shafts which give the crater-like effect of a qanat landscape (Fig. 68).

The origins of qanat are very ancient and probably date right back to the end of the second millennium B.C., while some of the most sensational, reputedly with headwaters up to 100 m deep and a length of 50 km, were probably built by the ancient Persians in the millennium before the coming of Islam in the seventh century A.D. The diffusion of this technique from a possible point of origin in the ancient mining areas to the north-west of Persia into Central Asia, Baluchistan, Arabia, the Sahara and the Mediterranean world, and the reasons for its appearance in such unexpected places as Bavaria, Liège, Madrid, and even South America, are complex and not fully understood. However, it emphasises the necessity for studying historical factors when trying to understand why any of the techniques described in this chapter may be found in certain areas and not in others to which they are environmentally suited.

Another basic lesson that can be learned from a study of the qanat system is that it is not only technique that is important in determining how and where man can live in the desert. Organisational ability is of prime importance and if political and social organisation breaks down, so will the water utilisation systems. The development of marginal land by the qanat system, for example, would never have been possible without the kind of political and economic security that the Achaemanids, the first great line of Persian monarchs, provided after their rise to power some 2500 years ago: 'they granted the enjoyment of the profits of the lands to the inhabitants of some of the waterless districts for five generations on the conditions of their bringing freshwater in' (Polybius: X, 28).

Artificial lift

Whereas qanat use gravity flow to bring the water to the place of exploitation, artificial lift has to be employed for bringing groundwater to the surface in flat areas. Once again a wide range of well devices is traditionally used (Fig. 69): most of them are now abandoned for tube wells and the ubiquitous diesel pump. In India, a colossal electrification programme has enabled many villagers to use electric pumps.

However, it is not only in the exploitation of groundwater that the problem of lift is met. It also arises on rivers, particularly in places where the channel is incised or where the water level varies seasonally. It is here perhaps that there has been the most change due to modern technical innovations. For example, during most of its history water use on the Nile has largely been confined to exploiting the flood by means of basinal irrigation. It was not until the nineteenth century that base flow was utilised for perennial irrigation thanks to the construction of barrages: and it is only in the twentieth century that the flood has been fully utilised and irregu-

69 Although rural electrification has been greatly extended in India in recent years, oxen are still used in many villages to lift water in leather containers from shallow wells. The well in this picture is surrounded by a small fence of succulent and prickly Euphorbia, and the water is poured into small channels which lead the water onto the cultivated area.

larities in the regime evened out through the construction of storage dams, of which the Aswan High Dam is the most notable example.

The future

Just as technical innovations combined with man's organisational capacity explain much about past settlement of the arid lands, so they hold the key to the future. Miracles, however, should not be expected. On the one hand the potential for discovering new hydrological reserves is limited, while on the other desalination remains dogged by exactly the same problem as has always faced it ever since the principles of distillation were first discovered – the need for a cheap energy supply.

At present it seems that the likelihood of discovering such a supply is remote. The prospects offered by processes that desalinate by distillation are very limited and most progress is being made in improving engineering design for existing systems. Some processes which use ion-selective membranes, notably electrodialysis and reverse osmosis, are now being used for the up-grading of brackish water and experimental work is being carried out on crystallisation (by freezing) for precipitating salts. Possibilities of directly using the sun's energy in solar stills are beset by the basic problem of concentrating incoming radiation. Something like one square metre of glass or plastic cover over the absorbing black surface is required to produce a litre of condensation per day: such a supply may be useful for drinking water in isolated communities and perhaps a small production of salad vegetables for a wealthy urban settlement, but little

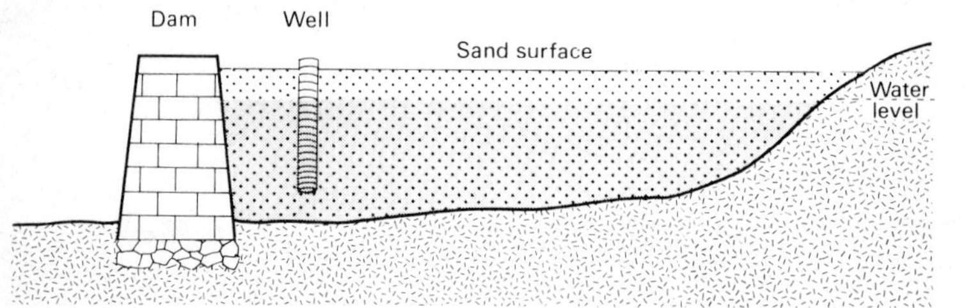

70 A sand- or gravel-filled reservoir. The water level is kept below the sand surface level to shield it from evaporation. Although water storage volume is reduced by over 50%, evaporation may be reduced by 90%. The water is obtained by means of a well.

else. With all these techniques costs per unit volume of water are still very high and at present can only really be justified for limited urban use and for settlements in areas where fresh water is simply not available. The price is certainly far too high for extensive agriculture, though there is perhaps some potential in areas of low-energy costs for mixing desalinated waters with ground-water when this is only marginally too saline (e.g. in Eastern Province of Saudi Arabia).

Some progress is being made in the fight against salt through the results of both fundamental and empirical research in two fields. One is in the development of salt-tolerant strains of crops, and the other is in developing better water use and drainage so that salinity does not build up. Experiments in irrigation with brackish water in well-drained soils are also producing interesting results particularly with trees and other plants that have a potential environmental and grazing value.

Hopes of increasing the natural supply of freshwater must remain limited, but here the problem can be attacked on a wider front; through exploration, through improving the efficiency of water use, and through increasing the yields of present sources of supply.

Improved methods of water conservation, though often unspectacular, may also do much to improve water supplies. Reduction of water loss from evaporation by such techniques as the sand-filled reservoir used in South West Africa (Namibia) (see Fig. 70) or by using a chemical, wax or plastic coating may prove successful. Seepage losses from canals, reservoirs and highly permeable soils may be reduced by simple techniques like soil compaction. Mulches, whether of plant debris or of plastic sheeting can reduce evaporation from soil surfaces. The degree of transpiration from plants may be reduced by a variety of means, including the destruction of unwanted phreatophytes, the removal of unproductive leaves, the breeding of strains which transpire less, and through the use of chemical anti-transpirants, which either promote stomata closure, or form a film over leaves. The wastage of water during irrigation can, likewise, be reduced by the adoption of trickle or drip irrigation. Standard methods are particularly wasteful because the extended areas of wetted soil greatly encourage evaporation. Trickle or drip irrigation uses a system of plastic pipes which drip water into the soil through outlets arranged near each individual plant (Fig. 71).

Hydrological prospecting will certainly continue to reveal reserves of freshwater in desert areas and will improve our knowledge of existing aquifers. But because the absolute quantity that can exist is limited, for reasons which have already been discussed, the probability is that new discoveries will be of decreasing

71 In Bahrain, a government experimental station has been watering tree crops with trickle irrigation from plastic piping. This localised application of water is much less wasteful than traditional methods which tend to apply water over the whole ground surface, thereby causing high evaporation losses.

economic value. Even now fairly extensive proven water resources in the arid portions of some underdeveloped countries are not being exploited because the cost of developing them for a sparse nomadic population dispersed in remote areas is prohibitive.

So for the foreseeable future development of the arid lands is going to stem chiefly from undramatic but steady improvements in present methods of exploitation and in the efficiency with which water is used. Harnessing desert rivers for both their irrigation and hydro-electric potential may, of course, still produce schemes that make headline news, but much of the real progress will be made in the less sensational fields of regional co-ordination and integrated economic effort, along with social and land reform in the riparian states. Similarly a technological breakthrough in producing an effective monomolecular film for covering exposed water surfaces so as to cut down evaporation for great reservoirs like Lake Nasser will make news, but less attention will be drawn to the task of changing cultivation methods so as to reduce over-irrigation and other wasteful use of hard-won water.

Economic return will also come from optimising the use of water in another way. In agriculture the tendency often is to consider how the yield of one plant may be increased, but this is not necessarily the most profitable line of research where water is the limiting input: it may be that fewer waterings over a wider area will result in greater total production than would be the case if water were applied to maximise the yield of each individual plant.

Table 12. The areas of Egypt under cultivation with selected crops

	Agricultural year	
	1951/2	1970/1
Cotton	1967	1525
Rice	362	1135
Maize (summer)	27	1171
Maize (winter)	1677	351
Sugar-cane	92	193

Areas are expressed as feddans x 10^3. 1 feddan = 4208 m^2 (1.038 acres).
Source: *Statistical Abstract of the Arab Republic of Egypt 1951/2* and *1970/1.*

It was partly on this basis, for example, that the norm of water application for growing clover was reduced from 5000 m^3/ha in 1955 to 3000 m^3/ha by 1961 on the Israeli coastal plain. Optimising the agricultural yield from a given input of water may also involve modifying rotations and a switch of crops to meet changing economic circumstances. Since the completion of the Aswan High Dam new problems of land allocation have arisen, for now it is not basically a water shortage that faces Egypt but an insufficiency of high grade soils to use it on. At the same time the pressure on agricultural production to feed her rapidly expanding population, to provide raw materials for industry and to meet her growing import bill has increased. So the combination of new economic requirements coupled with an increased water supply for summer irrigation has led to a major shift in the season for growing maize, (with a corresponding 75% rise in yield per unit area), an enormous increase in sugar production from Upper Egypt (partly at the expense of cotton) and a new concentration on rice growing, largely for export (see Table 12).

But it is not only in agricultural production that water use has to be optimized. Urban and industrial requirements are increasingly entering into competition with agriculture for this scarce resource. Such a situation has for long existed in Israel, where the total annual water supply is only in the order of 1½ milliard m^3, but today it is becoming of increasing concern for a rapidly developing country like Iran, notably in the main areas of demographic and economic transformation. Around Tehran, for example, the traditional qanat (see pp. 75–6) supply of roughly 40 million m^3 per annum has already had to be increased to something like six times that figure, despite a policy of limiting industrial licences, whilst in Isfahan, which is becoming a centre for major industrial development based on the Aryamehr steel mill complex, the supply of the Zayandeh Rud (the 'life-giving river') is already inadequate to meet the needs of both agriculture and industry.

In conclusion, therefore, it is better to take a pessimistic rather than an optimistic view in the medium term, and to recognise the problems of the arid lands rather than to engage in wishful thinking. Spectacular advances may occur in some of the developed countries and the new oil-rich states, but this is because they have money to invest in long-term research and in projects of dubious short-term economic value. Perhaps a return of more general value will stem from this experience but it is doubtful whether the vast Sahelian zone, for example, will benefit much in the foreseeable future. Indeed the general trend on the world scale would seem to be for such areas of marginal land use to become increasingly backward and for the frontier of settlement in arid lands to withdraw rather than to advance.

READING

The literature on deserts is now vast. Cooke & Warren, in their book on desert landforms, cite a multilingual literature of about one thousand books and papers.

Good general treatments of desert landforms occur in:
J. Tricart & A. Cailleux (1969) *Le modèle des régions sèches* (Centre de Documentation Universitaire, Paris),
and in
R.U. Cooke & A. Warren (1973) *Geomorphology in deserts* (Batsford, London).

Desert vegetation and animal life have been analysed by:
K. Schmidt-Nielsen (1964) *Desert animals: physiological problems of heat and water* (Clarendon Press, Oxford),
J.L. Cloudsley-Thompson & M.J. Chadwick (1964) *Life in deserts* (Foulis, London),
and
H. Walter (1971) *Ecology of tropical and subtropical vegetation* (Oliver and Boyd, Edinburgh).

Useful general reviews over a wider range of desert environments include:
J. Demangeot (1972) *Les milieux naturels désertiques* (Centre de Documentation Universitaire, Paris),
and
W.G. McGinnies, B.J. Goldman & P. Paylore (1969) (eds), *Deserts of the world* (Arizona University Press, Tucson).

A book which emphasises the special characteristics of coastal deserts is:
P. Meigs (1966) *Geography of coastal deserts* (UNESCO, Paris).

A general book which blends physical and human geography together and gives a number of apposite regional examples is:
Dov Nir (1974) *The semi-arid world* (Longmans, Harlow).

There are also many specifically regional studies, as for example:
are:
M. Evenari, L. Shanan & N. Tadmor (1971) *The Negev: the challenge of a desert* (Harvard University Press, Cambridge, Mass.).
and
J.C. Wilkinson (1977) *Water and settlement in S.E. Arabia* (Clarendon Press, Oxford).

It would be impossible to give a comprehensive list of useful papers on deserts, but the following are valuable modern reviews which provide further bibliographic assistance on landforms, processes, hydrology and palaeoclimates:

R.F. Peel (1966) The landscape in aridity. *Transactions of the Institute of British Geographers,* **38**, 1-23.

R.F. Peel, R.U. Cooke & A. Warren (1974) The study of desert geomorphology. *Geography,* **59**, 121-38.

A.T. Grove & A. Warren (1968) Quaternary landforms and climate on the south side of the Sahara. *Geographical Journal,* **134**, 194-208.

A general review of desert climatology is:

R.D. Thompson (1975) The climatology of the arid world. *Geographical Papers No. 35.* Department of Geography, University of Reading.

A particularly comprehensive and well-referenced review of the potential for increased water use in deserts is:

National Academy of Sciences (1974) *More water for arid lands* (National Academy of Sciences, Washington DC).

GLOSSARY

Aggrade. To build up a surface with detritus, rock waste, sand, alluvium, etc. The opposite of degrade.
Allogenic means originating at a distance. In the case of streams it describes those deriving much of their water from afar.
Aquiclude is a formation which only absorbs water slowly and will not transmit it fast enough to furnish an appreciable supply for a well or spring. It is impervious to water moved by gravity.
Aquifer is a water-bearing bed or stratum.
Areic signifies an area without surface flow.
Artesian refers to any deep well, where because of a favourable arrangement of strata, water rises under pressure.
Aureole. A ring or zone surrounding the central part of a feature.
Bajadas (Bahadas) are formed where alluvial fans become laterally confluent and thus produce a continuous apron of waste bordering a mountain front.
Barchans are crescent-shaped dunes, the points of which are oriented towards the direction of movement.
Base-level. The level below which a land surface cannot be reduced by running water. The general and ultimate base-level for the land surface is sea level, but other local or temporary base-levels may exist, such as playa basins.
Biomass (of plants) can be defined as the total amount of living matter of plants above ground and underground in a given area.
Calcrete is a calcareous crust formed in the zone of weathering by accumulation of calcium carbonate (lime) in rock, soil or weathered material.
Creep is the gradual movement downhill of loose rock or soil due to alternate freezing and thawing, wetting and drying, or other causes.
Dayas are small, shallow hollows, occurring in desert limestone surfaces and probably formed by solutional processes.
Duricrusts are surface accumulations of iron, aluminium, silica, lime and other substances which may harden to given an indurated crust. They include laterites and calcretes.
Eluviation is the process by which material is removed in solution or suspension from the upper horizon or horizons of a soil.
Endoreic refers to rivers which flow into inland basins rather than into the sea. The opposite of exoreic.
Erg is a tract of sandy desert.
Evaporation is the change of water into vapour.
Evaporites are sediments which are deposited from aqueous solution as a result of evaporation of the solvent.
Evapotranspiration is a term embracing that portion of the precipitation returned to the air through direct evaporation or by transpiration of vegetation.
Exfoliation is the breaking or peeling-off of scales as concentric sheets from bare rock.
Gilgai is an Australian term for puff-like mounds developed in clay soils.
Gypcrete is broadly comparable to calcrete but involves the formation of a crust of gypsum (calcium sulphate).
Horizons are layers of soil approximately parallel to the land surface with more or less well-defined characteristics that have been produced through the operation of soil-building processes.
Hydrographs are plots of river discharge against time (see **Regime**).
Infiltration capacity. Water can only infiltrate into soil at a given rate. Infiltration capacity is a measurement of this rate.
Inselberg. An isolated hill.
Insolation. In the context of rock weathering it means the process of heating and cooling which may cause rocks to expand and contract so that they split.
Karstic is a term applied to areas of relatively soluble rock such as limestone that possess a topography peculiar to and dependent upon

underground solution and the diversion of surface waters to underground routes.

Leaching is the action of percolating water through soil and rock removing soluble constituents.

Lunettes are crescent-shaped dunes formed on the leeward sides of desert depressions.

Niche. A small environmental site where an individual organism can exist and develop.

Orographic rain is rain produced by uplift of air bodies as a result of upstanding relief.

Pan. A South African term for a closed depression.

Pediments are eroded bedrock platforms at the foot of a more abrupt hill slope.

Phreatophytes are plants which obtain their moisture from groundwater, often by sending down deep roots.

Piedmont. Literally 'at the foot of a mountain'. In desert areas it means low-angle surfaces both of transport (the pediment) and deposition beyond the mountain front.

Piezometric surface is an imaginary surface that coincides with the static level of water in an aquifer, to which the water will rise under its full head.

Playas are shallow temporary lakes or desiccated lake basins which are often enclosed and salty.

Pluvial refers to a period when rainfall was greater than it is now.

Regime is the general temporal pattern of hydrological events such as the pattern of river discharge during a year.

Rhourds are mountains of dune sand, often with a star-shaped or pyramidal form.

Sabkhas are salt-encrusted flats, especially coastal.

Siefs are longitudinal dunes, formed parallel to the wind and named after the Arabic word for sword.

Solifluction (or solifluxion). Slow flow downslope of rock debris saturated with water and not confined to definite channels.

Stone pavement is a surface material in which coarse particles are predominant at the surface but which is developed over a sediment with a larger proportion of fine particles.

Tafonis are small and large recesses in rock faces, resulting from differential weathering.

Transpiration is the emission of water vapour by plants.

Tufa is a porous formation of calcium carbonate, deposited around a spring.

Ventifacts are stones shaped by the wind. Those with three distinct facets are called **Dreikanter.**

Wadis are desert water courses.

Xerophytes are drought-resistant plants, i.e. plants structurally adapted for growth with a limited water supply.

Yardangs are wind-moulded ridges and grooves.

Zeugen are tabular rock masses on a pedestal of less resistant material which have been undercut by processes including wind abrasion.

INDEX